Asymmetric Killing

Asymmetric Killing

Risk Avoidance, Just War, and the Warrior Ethos

NEIL C. RENIC

OXFORD
UNIVERSITY PRESS

OXFORD
UNIVERSITY PRESS

Great Clarendon Street, Oxford, OX2 6DP,
United Kingdom

Oxford University Press is a department of the University of Oxford.
It furthers the University's objective of excellence in research, scholarship,
and education by publishing worldwide. Oxford is a registered trade mark of
Oxford University Press in the UK and in certain other countries

First Edition published in 2020

Published in the United States of America by Oxford University Press
198 Madison Avenue, New York, NY 10016, United States of America

British Library Cataloguing in Publication Data
Data available

Library of Congress Control Number: 2020934657

ISBN 978–0–19–885146–2

For Krista, my favourite person

Preface

Asymmetric Killing is concerned with war, and more specifically, the killing and dying that constitutes it. It examines the licence of combatants to harm and their liability to be harmed, as well as the historical and contemporary frameworks that underpin and justify this designation. Most importantly, it determines the extent to which the coherence of these same frameworks is imperilled by the capability of certain states to kill with little or no physical risk to their own forces.

Today, this mitigation of battlefield risk has culminated in the armed unmanned aerial vehicle (UAV), and specifically, its use by the United States in the absence of a significant military ground presence. What are we to make of the radically asymmetric nature of this violence? Is it ignoble? Is it immoral? Can it even be understood as war? Or should such concerns be dismissed as merely the latest misguided attempt to impose a quixotic standard of 'fairness' upon the battlefield.

Asymmetric Killing engages this debate, examining whether, and in what ways, the existing moral frameworks of war are challenged by the emergence of radically asymmetric violence. Readers of this book will gain a greater understanding of the indirect but critical role of reciprocal physical risk in the moral justifications for lethal force.

There exists a crossover point in war, at which the violence of the stronger party may be so one-directional as to undermine the moral basis for its permissiveness. As this book will illustrate, this point has been reached with the advent of UAV-exclusive violence.

Acknowledgements

This book would not have been possible without the steadfast support and guidance of my two PhD supervisors, Sebastian Kaempf and Christian Reus-Smit. When I first met Seb as a lowly undergraduate, he immediately sparked my interest in matters of war and peace. He saw enough in me to agree to act as my supervisor, for both Honours and PhD. During this time, I couldn't have asked for a more patient and dedicated mentor. Thank you, from a grateful friend. Chris has also been invaluable, educating me on both the importance and technique of crafting an argument. He has made me a better scholar, and continued to offer support as I navigate the challenges of early-career academia. For my part, I think the three of us make a fantastic team and I am immensely grateful to them both.

I would also like to thank the University of Queensland, and specifically, the School of Political Science and International Studies. I was older than most when I began as an undergraduate student. Old enough to fear that I was wasting my time; that higher education was beyond my limits. The speed with which I overcame this is owed in large measure to the real sense of community fostered within the department. My thanks go to the supremely talented and ever-friendly scholars who occupy it; it has been an absolute pleasure working alongside each of you. I'm especially thankful to the wonderful History and Theory group at the University of Queensland, and in addition to Chris and Seb, I thank Lorenzo Cello, David Duriesmith, Jacinta O'Hagan, Sarah Percy, and Andrew Phillips.

Parts of this book were completed at the University of Oxford, where I attended as a visiting scholar. It was an unforgettable fifteen months and I offer my thanks to all those who helped make it so. Special appreciation must go to Janina Dill and Henry Shue, whose ongoing input helped steer my project in a new and interesting direction. I would also like to thank Dominic Byatt and Matthew Williams at Oxford University Press for their guidance, diligence, and patience. It has been a real pleasure working with you both. I'm further grateful to the anonymous reviewers of this book for their insightful and constructive feedback.

Many friends also deserve gratitude for helping make this journey a special one. In Brisbane, thanks to all those who kept me focused: Ashleigh, Ryan,

Megan, Josie, Andrew, and Eglantine; as well as those who kept me distracted: Darcy, Schaller, Chris, Richard, Cameron, and Daniel.

I want to express my deepest gratitude to my parents for helping me realize my potential. My dad is the hardest working person I've ever met. It has been his example that I've looked towards when putting in the long hours necessary to complete this book. My lovely mum has helped keep me sane. I am particularly grateful for her open invitation to join her as a dog minder if academia doesn't work out.

Above all, thank you to my darling Krista. As I write this, we are about to embark on a new adventure in Germany. There is no one in this world that I would rather have at my side. Your love and support has meant everything, and it is to you that this book is dedicated.

Aristotle was on to something when he wrote that 'the roots of education are bitter, but the fruit is sweet'. To all those mentioned, thank you for making the roots a little less bitter and the fruit a little more sweet.

* * *

Some sections of this book have been published previously and I would like to thank the publishers for granting me copyright permission to use this material in this monograph. In particular, some parts of Chapter 2 and Chapter 7 have been published in N. C. Renic (2018), 'A Gardener's Vision: UAVs and the Dehumanisation of Violence'. *Survival* 60 (6): 57–72. Copyright © The International Institute for Strategic Studies reprinted by permission of Taylor & Francis Ltd, http://www.tandfonline.com/ on behalf of The International Institute for Strategic Studies. Some sections of Chapter 3 and Chapter 7 have been published in N. C. Renic (2019), 'UAVs and the End of Heroism? Historicising the Ethical Challenge of Asymmetric Violence'. *Journal of Military Ethics*, published online 19 March 2019. DOI: 10.1080/15027570.2019.1585621. https://www.tandfonline.com/. Finally, some parts of Chapter 4 and Chapter 7 have been published in N. C. Renic (2019), 'Justified Killing in an Age of Radically Asymmetric Warfare'. *European Journal of International Relations* 25 (2): 408–30. Copyright © 2019 by the Author. Reprinted by permission of SAGE Publications, Ltd.

Contents

List of Abbreviations

API	1977 Additional Protocol I to the Geneva Conventions
AQAP	Al-Qaeda in the Arabian Peninsula
BCE	Before Common Era
Brig. Gen.	Brigadier General
Capt.	Captain
CCW	Convention on Prohibitions or Restrictions on the Use of Certain Conventional Weapons
CIA	Central Intelligence Agency
COIN	Counterinsurgency Operations
Col.	Colonel
DASH	QH-50 Drone Antisubmarine Helicopter
DDE	Doctrine of Double Effect
DFC	Distinguished Flying Cross
DoD	Department of Defense (US)
DPT	Democratic Peace Theory
FOFA	Follow-On Forces Attack
FRY	Federal Republic of Yugoslavia
Gen.	General
ICJ	International Court of Justice
ICRC	International Committee of the Red Cross
ICTY	International Criminal Tribunal for the former Yugoslavia
IDF	Israeli Defence Forces
IHL	International Humanitarian Law
IHRL	International Human Rights Law
INGO	International Non-Governmental Organization
ISAF	International Assistance Security Force
ISIS	Islamic State in Iraq and Syria
JSOC	Joint Special Operations Command
LOW	Laws of War
Lt. Col.	Lieutenant Colonel
Maj.	Major
Maj. Gen.	Major General
MoD	Ministry of Defence
MON	Memorandum of Notification
NATO	North Atlantic Treaty Organization
NGO	Non-Governmental Organization

NIAC	Non-International Armed Conflict
OAF	Operation Allied Force
ODS	Operation Desert Storm
Pvt.	Private
PTSD	Post-Traumatic Stress Disorder
RAF	Royal Air Force
ROE	Wartime Rules of Engagement
RPV	Remotely Piloted Vehicle
SEAL	Sea, Air, and Land
UAV	Unmanned Aerial Vehicle
UCLASS	Unmanned Carrier-Launched Airborne Surveillance and Strike System
US	United States
USAAF	United States Army Air Force
USAF	United States Air Force

Introduction: The Moral Limits of Asymmetric Killing

Ernst Jünger volunteered for the German Army on 1 August 1914, at the age of 19. For the next four years he fought on the Western Front, until wounded for the fourteenth and final time by a British bullet to the chest. Jünger survived and in 1920 published his memoir, titled *Storm of Steel.* Few works before or since have more effectively captured the experience of modern combat. The following is an account of the 1917 Battle of Cambrai, from the perspective of Jünger:

> [Y]ou hurled your own bomb, and leaped forward. One barely glanced at the crumpled body of one's opponent; he was finished, and a new duel was commencing. The exchange of hand-grenades reminded me of fencing with foils; you needed to jump and stretch, almost as in a ballet. It's the deadliest of duels, as it invariably ends with one or other of the participants being blown to smithereens. Or both. (2004: 214)

Jünger, though himself a famously courageous and decorated soldier, offers a fairly representative description here of close quarter battle. Grenades aside, his portrayal of the violent and frenzied reality of combat would have been equally well suited to most of the battlefields of history. The same cannot be said of the following example. In a 2009 article in *The New Yorker*, Jane Mayer recounted the lethal targeting by American forces of then Pakistan Taliban leader Baitullah Mehsud:

> It was a hot summer night, and... [Mehsud]... was joined outside by his wife and his uncle, a medic; at one point, the remarkably crisp images showed that Mehsud, who suffered from diabetes and a kidney ailment, was receiving an intravenous drip. The video was being captured by the infrared camera of a Predator drone, a remotely controlled, unmanned plane that had been hovering, undetected, two miles or so above the house... The image remained just as stable when the C.I.A. remotely launched two

Asymmetric Killing: Risk Avoidance, Just War, and the Warrior Ethos. Neil C. Renic, Oxford University Press (2020).

DOI: 10.1093/oso/9780198851462.001.0001

> Hellfire missiles from the Predator. Authorities watched the fiery blast in real time. After the dust cloud dissipated, all that remained of Mehsud was a detached torso. Eleven others died: his wife, his father-in-law, his mother-in-law, a lieutenant, and seven bodyguards. (2009)

It is difficult to imagine two more contrasting episodes of military violence. Consider first the issue of distance. Jünger provides a quintessential depiction of face-to-face combat, in which a soldier's projection of and exposure to violence is measured in metres, if not inches. This close proximity is absent in the killing of Mehsud. Two miles separated the target from the robot that killed him—potentially tens of thousands of miles if we factor in the American operator who triggered the attack. Both cases are also temporally distinct. There is a furious intensity to Jünger's battlefield that serves to narrow the scope of decision-making. No sooner had one life-threatening encounter been resolved than another emerged. In contrast, the capability of unmanned aerial vehicles (UAVs) to loiter over potential targets affords the operator hours, if not days, of cool deliberation before a decision is made to terminate.

Arguably nothing, however, separates these two episodes to a greater degree than the issue of physical risk. For Jünger, just as for a majority of those who have participated in war throughout history, the risk of injury and death in battle was an inescapable condition. The Mehsud killing is a paradigmatic example of the extent to which this long-standing feature of war has concluded, at least for the stronger party, within certain contemporary contexts. What has emerged in its stead is radically asymmetric violence—lethal force actualized by one side in the near-complete absence of physical risk, across the full scope of a conflict zone.[1] The focus of this book is whether this intensifying risk imbalance represents a challenge to the existing moral frameworks that justify killing in war.

Obtaining a full and accurate account of the moral significance of radical asymmetry is of both theoretical and practical value. For those who participate in war there are few questions more important than precisely when is it permissible to kill? 'Lethality', wrote US General (Gen.), Martin E. Dempsey, 'is the foundation on which everything we do must be built, but lethality brings with it incredible obligations and responsibility' (2010: 154). This

[1] The term 'asymmetrical warfare' is most commonly used to describe the unconventional tactics leveraged by weaker parties to expose the vulnerabilities of a militarily superior opponent. This mode of warfare is directed in such a way as to undermine the stronger parties' capacity and will to fight on (McKenzie 2000: 2). To be clear though, the term 'asymmetry' is employed here in reference to disparities of *physical* risk only.

includes the need to reflect upon, and if necessary, re-evaluate the theoretical basis for this licence. The following study will stand as one such assessment. Importantly though, this book will not set out its own normative theory. Instead, it situates the phenomenon of radical asymmetry within the morally contested terrain of the regulatory frameworks of war, to illustrate its tension with the existing justifications for lethal force. It seeks to identify whether disparities of physical risk between opponents in war may reach such a degree as to render ambiguous what would otherwise be a morally unproblematic use of military violence.

Establishing the significance of this asymmetry is not merely a theoretical exercise, but also of practical value, given the pre-eminence of the American military. The United States, through the application of technocentric and casualty-averse modes of warfare, is increasingly able, within certain conflict zones at least, to kill without any recognizable element of physical risk to its own forces. An early glimpse of this asymmetry could be witnessed during the 1999 US-led NATO intervention into Kosovo—the first zero combat casualty war in military history.[2] Since then, the capability of the United States to apply one-directional force to the battlefield has only increased, most notably through its growing dependence on military robotics. Today, this military overmatch has culminated with the UAV, and specifically, its utilization in the absence of a significant ground presence in areas such as Pakistan, Yemen, and Somalia. It is estimated that between November 2002 and June 2014, 98 per cent of the targeted killings conducted by American forces in these theatres were executed via armed UAVs. The remaining 2 per cent of strikes were comprised of armed aircraft attacks and ground raids (Zenko and Kreps 2014: 9–10). The ability of the United States to physically dislocate its own human combatants from the battlefield and deliver UAV-exclusive violence from up to a hemisphere away goes beyond mere advantage, conveying a military dominance that approaches impunity. Establishing the degree to which this dominance represents a challenge to the moral permissibility of killing in war will be the focus of this book.

Any inquiry into the moral permissibility of killing in war must begin with the Just War Tradition, and specifically, the framework of *jus in bello*, the moral rules that govern battlefield conduct once hostilities have commenced.[3]

[2] While the US did not experience a single combat casualty, two American servicemen were killed during the course of the conflict, when their AH-64 Apache helicopter crashed while conducting a night-time training exercise (Graham and Drozdiak 1999: A1).

[3] *Jus in bello* functions at the midway point between *jus ad bellum*, the set of criteria consulted to determine whether entering a war is morally permissible, and *jus post bellum*, the rules that apply at the termination phase of war.

Analysis should not, however, be restricted exclusively to these explicit rules and standards. Of equal importance are the deeper assumptions that underpin this moral framework, as well as that of war itself. Consider that even when heavily regulated, the scope and intensity of violence that may be applied and received between combatants in war goes far beyond what would be acceptable in virtually any other context of human interaction. What are the assumptions that inform this permissiveness? And just as importantly, is there a point at which the material structure of war may shift to such a degree as to render these assumptions untenable? Establishing the challenge of radical asymmetry is contingent on more than a finding of rule conformity or deviation by relevant actors. It also demands an investigation of the indeterminacy that may result when these moral rules become unmoored from their deeper logical premise. This book will provide such an account, establishing the extent to which the moral foundations that justify killing can maintain their coherence within the context of increasingly one-directional violence.

In summarizing this section, it is important to clarify that 'radical asymmetry' is not synonymous with 'risk-free' war. Despite the military capacity of the United States, total and continuous immunity from physical vulnerability in war cannot be established at present, a reality that is likely to endure for the foreseeable future. For this reason, true risk-free warfare is best understood as a Weberian 'ideal-type'.[4] Radical asymmetry does, however, describe a battlefield condition in which one side has, in a practical sense, overcome the type and degree of physical vulnerability that has long been regarded as an expected and unavoidable feature of war.

The traditional battlefield is both punctuated and pervaded by violence. Combatants are most obviously in danger when they directly engage those actively seeking their destruction. Injury and death is a potential, sometimes even likely, outcome of these moments. Crucially though, even outside these sporadic encounters, some degree of danger endures. In most circumstances, quotidian acts such as showering, lighting a cigarette, eating, and sleeping, can be undertaken with a virtual guarantee of safety. Not so on the battlefield where even the most innocuous activity can be a precursor to violent harm. Radical asymmetry is obtained when one party divorces itself, entirely or near enough, from this sustained and pervasive imposition of danger. The surest

[4] For Weber, ideal-types, or 'thought pictures' (*Gedankenbilder*), were heuristic devices tasked with the 'analytical ordering of empirical social reality' (1946: 61). Put simply, ideal-types are conceptual tools that offer a 'pure form of a particular phenomenon to better enable comparative analysis' (Coker 2010: 13).

way to achieve this outcome is the physical removal of human combatants from the battlefield itself.

The distinctiveness of radical asymmetry is also a matter of scope. In the case of UAV-exclusive violence, an unprecedented degree of military imbalance has been imposed by one party, not merely in a specific battlefield encounter, but across entire conflict zones. The critical question is whether the scope and intensity of this imbalance constitutes a crossover point between morally unproblematic and problematic killing.

Asymmetry in War

To properly engage with this question, it should first be acknowledged that while UAV-exclusive violence represents the most contemporary and extreme manifestation of this trend, risk-asymmetry in war is far from a recent phenomenon. Battlefield assumptions have long been tested by the ability of one side to kill in the partial absence of physical risk. During the 1898 Battle of Omdurman, for instance, Maxim machine gun and artillery equipped British forces killed 11,000 Sudanese fighters and wounded another 16,000—all at a loss of forty-eight men (Ellis 1986: 86–7). British journalist G. W. Steevens, who accompanied General (Gen.) Kitchener throughout the campaign, described the encounter not as a 'battle, but an execution' (1898: 264). A younger Winston Churchill was also present and offered an equally stark account, heralding it as 'the most signal triumph ever gained by the arms of science over barbarians…[achieved]…with hardly any difficulty, comparatively small risk, and insignificant loss to the victors' (1902: 300).

Comparable military imbalance has emerged throughout the history of war as a result of a range of technological innovations. These include early projectile weapons, firearms, the submarine, military sniping, and manned aerial bombing. Of particular relevance to this book, each of these modes of violence, upon their introduction, was argued to distort the reciprocal imposition of risk between enemies to such a degree as to obscure the 'vital distinction between war and plain murder' (van Creveld 1989: 71). 'Underhand, unfair, and damned un-English' was the description of submarines first offered by Admiral of the British Royal Navy, A. K. Wilson (1902, cited in Marder 1961: 332). Wilson went on to advocate that captured crews of enemy submarines be hanged as pirates.

The challenges that emerged as a result of these disparities of risk were grounded not only in resentment by those incapable of replicating similar military feats, but also a real sense that such imbalance represented a battlefield transgression. This is evidenced by the frequency with which opposition to such asymmetry extended beyond those targeted, to include its beneficiaries. Crucially though, whatever the degree of initial resistance, these historical tensions have ultimately failed to endure. In the face of conceptual and material change to the character of war, expectations of legitimate conduct and military expediency have shifted to better accommodate these modes of asymmetric violence. The archer, the sniper, the submariner—despite sustained criticism of their conduct, each has undergone a gradual normalization of status.

The consistency with which these historical challenges have been overcome has reinforced an understandable consensus within much of the relevant academic literature; that beyond any passing ethical discomfort, asymmetries of threat and risk, no matter how profound, have little bearing on the *in bello* status of battlefield violence. As such, the permissibility of radically asymmetric killing is to be determined, just as with all other military force, on the basis of its conformity with the principles of discrimination and proportionality.

This logic is dominant within a majority of sources dealing with the emergence of radically asymmetric technologies, including the UAV and other types of robotic warfare (Brooks 2012a; Coady 2008a: 174; Lamb 2013; McMahan 2013: x; Strawser 2010: 343). When the asymmetric potential of such violence is challenged, it is most often done so by those concerned that such disparity is obtained at the expense of a reduced threshold for war (Anderson 2012; Asaro 2008: 7; Brooks 2012b; Brunstetter 2012; Chenoweth 2012; Freedman 2016: 153–8; Johansson 2011; Kaag and Kreps 2014; Mazzetti 2013: 6, 100; O'Connell 2011; Sauer and Schörnig 2012; Sparrow 2012: 60; Spetalnick 2014; UK MoD 2011: 55–6; Welsh 2015: 25), or alternatively, the safety of enemy non-combatants (Cook 2004: 127; Luttwak 1999: 41; Shaw 2005: 22; Shue 2003; Shue 2011: 137; Williams 2008: 589). The suggestion that radical asymmetry is a challenge to the permissibility of *intercombatant* violence is typically dismissed by these same sources as the product of an increasingly antiquated understanding of war as an equal contest.

What this dismissal too often overlooks is the extent to which the functionality of the rules and standards of war rests not only on actor compliance, but also more fundamentally, on the general acceptance that the frameworks themselves are coherent and non-arbitrary. This acceptance may erode, either when the rules put forward are imprecise, or in the event of a significant enough shift in the material conditions within which they operate. In either case—deficiency or obsolescence—the capacity of these frameworks to render clear moral judgements is reduced. In the context of radical asymmetry, there exists a need for sustained reflection of the extent to which the coherence of the moral justifications for killing rests upon *assumed* conditions of mutual threat and endangerment between antagonists. The failure of existing literature to theoretically and historically ground the principle of reciprocal risk within these frameworks has contributed to a range of claims regarding the challenge of radical asymmetry that are not yet credible. In particular, such scholarship currently underestimates the potential of this challenge to call into question the underlying rationale for the moral right to kill. This same scholarship has proven equally incapable of providing a satisfactory explanation of how critical a difference exists between the significant reduction of battlefield risk, observed throughout the history of war, and its complete or near-complete elimination by one side at the collective level.

The need for deeper analysis is further evident in relation to those sources that do consider the emergence of radical asymmetry a moral challenge (Enemark 2013a; Ignatieff 2000: 161; Kaempf 2014; Kahn 1999: 201; Kahn 2002: 2–7; Sparrow 2013). This literature provides a useful starting point to uncover the significance of profound risk-disparity in war. Like those sources that dismiss this challenge outright, however, this scholarship has yet to properly clarify both the explicit and tacit role of reciprocal risk within the moral frameworks that justify violence. This confirms the need for a systematic explanation of the function of risk in the right to kill, as well as our very understanding of what forms of violence constitute war.

Bridging the Research Gap

A lack of sustained analysis has so far hindered the development of a plausible account of both the challenge and implications of radical asymmetry for the morality of killing in war. This book will address this oversight through a historical study of the role of reciprocal risk in the ethical (warrior ethos) and moral (Just War Tradition) justifications for lethal force.

Before detailing this approach, it is important to specify how the terms 'ethic' and 'moral' will be employed and differentiated in the chapters ahead. According to Walker and Lovat, ethics are agent-centric; they focus on how one should live one's own life from the perspective of one's own values (2016: 436–7). The warrior ethos is understood as an *ethical* framework, primarily combatant-focused and more practical than abstract in nature. This is reinforced by Coker, who states: 'to be moral an action must be disinterested...most soldiers...live in a distinctive ethical community. Ethics are inherently interested' (2010: 187–8). The Just War Tradition, in contrast, has long represented a predominantly academic discourse. Its primary focus is on the potential targets, rather than agents of violence. This aligns with Harper's description of morality as 'other-regarding' (2009: 1066). Differentiating the term 'ethic' from 'moral' provides a lexicon for this book to explore and contrast between the obligations of the warrior ethos and Just War Tradition.

This, however, begs the obvious question of why include the warrior ethos at all, given the stated focus on the *moral* significance of radical asymmetry. The answer lies in the historical salience of this issue. As the previous section outlined, opposition has emerged throughout the history of war in response to a range of practices that intensified battlefield asymmetry. Crucially though, whatever the impact of this opposition at the time, it has ultimately failed to endure. Informed by this history, many contemporary scholars conclude that radically asymmetric violence fails to problematize the moral right to kill, and is instead merely the latest in a long trend of resolvable tensions with battlefield ethics. In order to test the validity of this claim, the moral *and* ethical frameworks of war must be examined.

It should be stressed that many authors draw the distinction between 'ethic' and 'moral' either differently, or not at all. To add yet another layer of complexity, the term 'Just War Tradition' is used by some to refer to all forms of battlefield regulation, including chivalric military ethics, secular moral judgement, ecclesiastical doctrine, and even the laws of war. Within this book, however, the warrior ethos is considered independently from the other components of Just War. This is partly done for the sake of methodological clarity. More importantly though, by separating these frameworks we are better able to identify key distinctions in logic. In his work on medieval Just War, Johnson notes that while chivalric (ethical) and canonical (moral) perspectives aligned regarding the importance of non-combatant immunity, the principles and assumptions that underpinned their support differed (1981: 138–40). This distinction endures, particularly in relation to the precise role of physical risk and material threat in the justification for violence.

This book will evaluate the extent to which—amidst the change and variance of the history of war—a thread of reciprocal risk has endured as an underpinning assumption in the right to kill, within both the warrior ethos and Just War Tradition. Just as importantly, it will identify the level at which this assumption operated. Do the frameworks that justify killing draw their coherence from the mutual risk of war, and crucially, if so, does that translate into an expectation of inter-combatant (individual) or inter-belligerent (collective) risk? In other words, is this expectation satisfied only through a direct imposition of physical risk between individual enemies at the moment of contact, or alternatively, is it sufficient to maintain a structural reciprocity of risk between warring belligerent parties? This approach will clarify the degree to which the core assumptions of both frameworks are challenged, and potentially undermined, by radically asymmetric violence.

There of course exists a legal dimension to each of these questions, and today the laws of war arguably function as the primary source for evaluating the conduct of fighters.[5] Despite this, the book does not examine the potential legal challenge of radical asymmetry. This decision is motivated partly by the need to set practical limits on the scope of the project. More importantly, an examination of the law is unlikely to yield sufficient insight regarding the disruptive potential of radically asymmetric violence. Put simply, there exists no formal law of war that restricts the right of combatants to kill only to situations in which they are personally subjected to physical risk. Combatants may be injured or killed by their opponent at any time, regardless of their actual threat to the enemy (Dinstein 2007: 144; Parks 1989; Sassòli and Olsen 2008: 605–6). This legal permissiveness is reflected in most operational guidelines for combatants. The US *Wartime Rules of Engagement* (ROE), for example, 'permit US forces to open fire upon all identified enemy targets, regardless of whether those targets represent actual, immediate threats' (US Army 2000).

It is the intention of this book, however, to go beyond these specific rules and examine the bedrock of ethical and moral assumptions that sustain their coherence. It seeks to establish whether radical asymmetry is a challenge *despite* the fact that it fails to explicitly violate the rules of war. In light of this emphasis, this book will engage with the diversity of opinion found within the warrior ethos and Just War Tradition—both their modern configurations and historical antecedents.

In addition to its study of the role of reciprocal risk in the assumptions of war, this book will be anchored within a historical account of the challenge of asymmetric violence. This will consist of an analysis of three cases: two historical episodes of asymmetry—military sniping and manned aerial bombing; and one contemporary case—UAV-exclusive violence. These cases were

[5] By the mid-nineteenth century, the Just War Tradition had undergone a transition from moral ideals and religious canon to a series of military laws established by the state to govern the conduct of combatants. From this period onwards, legal empiricism has occupied the dominant position within international law (Hongsheng 2006: 274, Smith 2002: 358; Solis 2010: 7).

selected on the basis that each illuminates a key historical moment in which emerging changes to technology and doctrine generated an asymmetry of risk between opponents so profound as to directly challenge the existing frameworks that justified killing in war. An understanding of the dynamics that determined the way in which these historical challenges developed, progressed, and eventually resolved, will allow this book to engage current debates concerning radical asymmetry in a more comprehensive, historically informed manner. Similarly, by investigating key points of divergence between historical and contemporary cases of asymmetry, the extent to which the latter represents a qualitatively new phenomenon can be determined. Establishing this is of particular importance when assessing the broader implications of radical asymmetry for the regulation of lethal force.

The Argument

Pursuing these lines of analysis in Chapters 1 and 2, *Asymmetric Killing* first clarifies that within the moral frameworks of war the right of individuals to injure and kill in battle is *not* contingent on an assumption of personal physical vulnerability; i.e. inter-combatant risk. It is largely on this basis that the challenge of radically asymmetric violence has been dismissed within the literature. What this dismissal fails to properly consider, however, is the significance of collective, *inter-belligerent* risk within these same frameworks. This book argues that the *coherence* of the existing moral justifications for lethal force is grounded in an *assumption* of war as a site of reciprocal *structural* risk between belligerents. While not a formal requirement itself, this structural reciprocity provides the tacit rationale for the permissiveness of the moral standards of war. Structural risk, it is thus claimed, is not merely an empirically descriptive feature of war, but also a key moral one. Chapter 4 illustrates this in greater detail, engaging the three most dominant accounts of Just War today—'conventionalism', 'revisionism', and 'contractarianism'. The erosion of structural reciprocity through radically asymmetric conditions of battle represents a serious and sustained challenge to the coherence of each of these justifications for killing in war.

Structural risk also stands as the critical point of delineation between historical and current iterations of this challenge. For all their battlefield utility, neither military sniping (Chapter 5) nor manned aerial bombing (Chapter 6) sufficiently undermined the reciprocity of risk imposed between belligerents at the collective level of war. As a consequence, the tension that emerged from these modes of violence was principally grounded in their alleged incompatibility with the warrior ethos. The historical and contemporary role of this more narrowly focused, *ethical* conception of reciprocal risk will be outlined in Chapter 3.

The exclusive use of UAVs in conflict theatres such as Somalia, Pakistan, and Yemen pushes combatant preservation close to its logical endpoint. In doing so, it generates a challenge that goes beyond the warrior ethos. As Chapter 7 illustrates, the physical dislocation by the United States of its own targetable combatants concludes the long-standing configuration of the battlefield as a realm of pervasive and mutual threat. This, it is argued, injects a problematic ambiguity into the moral justifications for American violence, specifically in terms of its permissiveness. This is not to argue that within this context lethal force is now impermissible. What is suggested though, is an erosion in the ability of the United States to establish a morally coherent basis for liability to harm, particularly when those targeted are lower-level fighters.

What then are the practical implications of this ambiguity? Beyond its potential to render less tenable the presumptions by which moral liability to harm is designated, this book identifies two problematic outcomes of this military imbalance. First: radical asymmetry has distorted the distinction between war and more punitive, judicial violence to a degree that has imperilled the battlefield rights of those targeted. Specifically, the increasing absence of reciprocal risk between the United States and its enemies has contributed to a dehumanization of the latter, as well as an expansion of violence to those of more questionable targetable status. Second, radical asymmetry has narrowed the scope of legitimate retaliation for the weaker side, and in the process, incentivized a redirection of violence away from the battlefield and towards non-combatant targets. Both outcomes place further pressure on the frameworks of war and make the maintenance of mutually acceptable and applicable limitations on violence more difficult.

The Outline

This book is divided into three parts: I. Theory (Chapters 1 and 2); II. Reciprocal Risk in War (Chapters 3 and 4); and III. The Historical and Contemporary Challenge of Asymmetry (Chapters 5–7).

Part I. Theory

Chapter 1 will review the literature that engages the issue of radical asymmetry, with a particular focus on gaps in the research. It argues that one of the most problematic of these is a lack of sufficient theoretical and historical perspective. This chapter will then outline the methodological response of this book. This will include a historical study of the principle of reciprocal risk in war, alongside a more specified analysis of the asymmetry-challenges of military sniping, manned aerial bombing, and UAV-exclusive violence.

Chapter 2 will summarize the main contentions of this book. This will commence with an examination of the enduring role of structural reciprocity in the nature of war, followed by its more specific role in the moral justifications for killing. This will conclude with an overview of the historical challenge of asymmetry. The chapter will then detail the theoretical and empirical challenge of radically asymmetric violence. It argues that in both significance and implications, the moral tension generated by this structural imbalance of risk is qualitatively distinct from previous episodes of asymmetry.

Part II. Reciprocal Risk in War

Chapter 3 will locate the role of reciprocal risk within the warrior ethos. It first argues that exposure to personal, physical risk has long been regarded as a key element in the ethos-based conception of legitimate violence. As will be further shown, however, the warrior ethos is an evolving framework; one that gives increasing consideration to factors such as restraint and professionalism in determinations of ethical status. The adaptive quality of this ethos is a key explanatory factor in the historical resolution of asymmetry-challenges.

Chapter 4 will engage the *jus in bello* component of the Just War Tradition, beginning with the 'conventionalist' account of the moral right to kill. It then considers the most recent challenge to this position in the form of Just War 'revisionism'. Following this, it will examine the more consequentialist approach of 'contractarianism'. These three positions differ in their justifications for lethal force in a number of important ways and this chapter will not attempt to resolve this debate. Nor will it overly delve into ongoing debates within each position. What it will demonstrate is the extent to which an assumption of war as a site of reciprocal structural risk plays an essential role in the permissiveness of inter-combatant violence within each. This chapter will then outline the tension between the coherence of these justifications and conditions of radical asymmetry.

Part III. The Historical and Contemporary Challenge of Asymmetry

Chapters 5 and 6 will explore the historical asymmetry-challenges of military sniping and manned aerial bombing respectively. These chapters will first highlight the degree to which both modes of violence mitigated the reciprocity of risk between combatant and target, as well as the ethical tension that emerged as a consequence. It will then be argued that the failure of military sniping, and to a lesser degree manned aerial bombing, to sufficiently undermine the structural reciprocity of war limited the scope of their asymmetry-challenge to the warrior ethos. These chapters will conclude with an account of the gradual ethical normalization of both modes of violence.

Chapter 7 will outline the extent to which UAV-exclusive violence both aligns with and diverges from the more resolvable asymmetry-challenges of the past. This chapter first acknowledges the considerable overlap between historical and current asymmetry in terms of their tension with the warrior ethos. On this basis, it is reasonable to conclude that those who operate UAVs will gradually undergo a similar normalization of status. Crucially though, this analogy to the past breaks down in regard to the moral justifications for killing in war. This chapter argues that the unprecedented capacity of UAV-exclusive military force to erode the structural reciprocity of war has undermined the ability of the United States to interpret and apply morally acceptable limits on its violence.

In the concluding chapter of this book, the main contentions of this study will be revisited. This will commence with an overview of the significance and implications of the moral challenge of radical asymmetry. The broader relevance of these findings for ongoing debates within the Just War Tradition and laws of war will then be detailed. This will include analysis of the likely impact of the widening gap between what is morally justifiable on the battlefield and what is the legally permissible. Lastly, this conclusion will assess how the challenge of radically asymmetric violence is most likely to develop from this point onwards.

Concluding Remarks

Before commencing Chapter 1, further clarification regarding the US-centric focus of this book is required. Despite the military impact of the UAV, radical asymmetry is not the result of any single innovation in weaponry. Equally significant has been the rise of casualty aversion as a principal objective in contemporary armed conflict, as well as the emergence of de-territorialized global warfare, in which weaponized robots increasingly function as proxies for human combatants. Crucially, both these factors find their clearest expression—both in military doctrine and practice—in contemporary American warfare. Thus, while the manufacture and use of UAVs has spread internationally, the scale and scope of this phenomenon continues to centre to a substantial degree on the United States.[6]

This is significant when determining the broader contribution of this study. Jackson defined 'great powers' as the 'very select group of states whose policies and actions can affect the course of international affairs' (2000: 173). Though some view the trajectory of the United States as one of relative decline, it remains the foremost of today's great powers, with a global status that affords it a freedom of action that few other states can replicate. This status makes legitimacy challenges generated by its preferred mode of warfare particularly relevant. Former director of Amnesty International USA's Security and Human Rights Program, Naureen Shah, criticized the Obama administration's normalization of the 'bizarre', in reference to what she perceived as the problematic scale and opacity of its UAV program (cited in Devereaux and Emmons 2016). Through its actions and policy, the United States is currently in the process of establishing a normative and legal template that will inform the future use of radically asymmetric wartime practices. Cole writes that when developing nations seek out precedents for the use of robotic technologies in war, 'Obama's record will be Exhibit A' (2016).

For these reasons, this project will maintain radically asymmetric US violence as its central focus, while at the same time acknowledging the increasingly global scope of this challenge. Such a reflection is long overdue. As Chapter 1 will next outline, though there is an extensive and growing body of work centred on the emergence of radical asymmetry, much of it fails to accurately capture the precise nature of its challenge to the moral justifications for killing in war.

[6] The United States also continues to possess an unquestionable lead in the design, development, and production of UAV weapons platforms (Gilli and Gilli 2016).

PART I
THEORY

1

Asymmetric Violence and the State of the Field

Introduction

The ability to wage certain modes of contemporary warfare in the absence or near-absence of physical vulnerability has provoked extensive debate regarding the implications of such advantage. A majority of these responses, however, have fallen short of the theoretical and historical depth of analysis necessary for a full and accurate account of the issue. This is particularly evident in relation to the potential tension between increasingly riskless warfare and the moral right to kill. This chapter explores these limitations in greater detail, providing a critical evaluation of the existing academic response to the emergence of radical asymmetry in war. It then outlines precisely how this research gap will be bridged in the chapters ahead.

This chapter begins by clarifying the status of this issue within the literature. It argues that criticism of radical asymmetry, when it does emerge, is most often limited to its potential negative impact on either the political threshold for the use of force or the safety of enemy non-combatants. The chapter then details why, in contrast, this same literature has largely dismissed the potential of radical asymmetry to challenge the moral permissibility of *inter-combatant* violence. Following this, the chapter explores the less extensive but growing body of research that *does* engage this challenge. Problematically, many of these arguments are underpinned by a theoretically narrow and historically underdetermined conception of the role of threat and risk in the moral permissibility of lethal force. The chapter concludes with an overview of the methodology used within this book to address these gaps in the research.

Casualty Aversion and Moral Hazard

Today, the United States possesses a level of military pre-eminence so significant as to ensure that virtually any war it is likely to engage in will be characterized

Asymmetric Killing: Risk Avoidance, Just War, and the Warrior Ethos. Neil C. Renic, Oxford University Press (2020).

DOI: 10.1093/oso/9780198851462.001.0001

by a significant differential of military power. This is not to argue that the United States is no longer involved in conflicts in which its own soldiers are killed, nor is it to present the story of contemporary American warfare as one of uninterrupted technological triumph. American casualties incurred in both Iraq and Afghanistan highlight both the limits of technology in the context of insurgency warfare, as well as the selective willingness of the United States to expend lives in the theatre of war. This does not alter the fact, however, that the United States has achieved a degree of force protection that is arguably without historical precedent. Before engaging with the problematic dimensions of this achievement, the ideational factors that have brought it about should be acknowledged. Radical asymmetry represents the culmination of not only technological advancement, but also a doctrinal and political commitment to casualty reduction in war.

A range of explanations has been offered for the increasingly pivotal role of casualty aversion in contemporary, and particularly Western warfare. According to some authors, this military trend is but one element of a deepening and widening commitment by states to risk avoidance, brought about by an increase in the living standards of postmodern societies (Beck 1999; Coker 2002a: 57–9). This echoes a long-standing contention among political philosophers, from Tacitus (2016), to Jean-Jacques Rousseau (1992) and Adam Smith (1978), that affluence and commerce would destroy the warrior spirit.[1] Others view casualty aversion as the outcome of gradual change within the international system and international law. According to Meron, the strengthening role of individual rights-based approaches has contributed to 'a general distaste for the waste of human life' (2006: 6).[2]

For some authors, the degree of casualty aversion we are currently witnessing is best explained by more contingent factors. In the case of the United States, defeat in Vietnam is frequently presented as a critical juncture in their subsequent prioritization of force protection in war (Boot 2002: 319; DeGroot 1999: 267; Luttwak 1995: 122; Record 2000; Shaw 2005: 79). The Vietnam War, it is argued, established the limits of what the American public was willing to endure in relation to combat deaths, a lesson that would imprint itself on future US military doctrine (Bacevich 2005: 65; Buley 2008: 63; Reid 2002: 45; Scales 2005: 9; Schulzinger 2006: 530). 'America now is not willing to take any casualties', argued Henry Kissinger in 1999, 'Vietnam

[1] See Feaver and Miller (2014) for more detail.

[2] See also, Dill (2015a: 99); and Finnemore and Sikkink (1998).

produced a whole new attitude' (cited in Lacquement 2004: 40).[3] This explanation borrows from democratic peace theory (DPT), and specifically, the belief that ongoing domestic support is essential for the prosecution of liberal wars (Shaw 2005; Turns 2014: 212). This same theory also features in one of the main criticisms levelled against radically asymmetric violence—its alleged distortion of the political threshold for war.

The Threshold Challenge

Arguably the two most impactful (and divergent) legacies of Alfred Nobel were his invention of dynamite and subsequent founding of the Nobel Peace Prize. In 1892, Nobel reflected on the former during a conversation with the founder of the European anti-war movement:

> Perhaps my factories will put an end to war even sooner than your congresses. On the day when two army corps may mutually annihilate each other in a second, probably all civilised nations will recoil with horror and disband their troops. (1892, cited in Tägil 2014)

This well-intentioned, but ultimately quixotic aspiration has been repeated throughout history in reference to the potential deterrent effect of emerging military technologies, including air power (Hugo 1864, cited in Purdon 2010), the machine gun (Maxim, cited in Coker 2014: 24), and atomic/nuclear weapons (Oppenheimer, cited in Rhodes 2005). For many, the primary challenge of radical asymmetry is located in the inverse of this maxim: the suggestion that as the costs of war diminish, including the loss or even risk of loss of human life, democratic leaders will be greater empowered and incentivized to wage war in the absence of public consultation and support. These concerns pre-date robotic violence. During both the 1991 First Gulf War and 1999 Operation Allied Force (OAF), ongoing domestic support for American involvement was seen as contingent on the relative lack of necessary sacrifice (Ignatieff 2000; McInnes 2002).

In the years since, this 'threshold problem' has been explored by a number of authors, who link the increasing risklessness of certain modes of contemporary

[3] More recent conflicts in Iraq and Afghanistan have reinforced the importance of casualty avoidance in the maintenance of public support for war. It was noted as early as 2005 that casualty for casualty, support for the Iraq War declined more precipitously than during the Vietnam War (Gelpi, Feaver, and Reifler 2005; Mueller 2004: 44).

warfare to an erosion of democratic restraints on the resort to force (Anderson 2012; Asaro 2008: 7; Brooks 2012b; Brunstetter 2012; Chenoweth 2012; Freedman 2016: 153–8; Johansson 2011; Kaag and Kreps 2014; Mazzetti 2013: 6, 100; O'Connell 2011; Sauer and Schörnig 2012; Sparrow 2012: 60; Spetalnick 2014; UK MoD 2011: 55–6; Welsh 2015: 25).[4] Today, these concerns have intensified in relation to the radically asymmetric potential of UAV violence. 'If there are unusually usable weapons in the arsenal', notes Wrage, 'there will be unusual pressures to use them' (2003, cited in Cole 2016). One of the more troubling outcomes of this pressure is a potential increase in the use of lethal force as a first, rather than last resort (Galliot 2012: 63).

This is a valid line of inquiry, and suggests a need for further study. Too often, however, such analysis serves to perpetuate a false impression that radical asymmetry is a challenge only to the *ad bellum* status of the stronger side. What is neglected is whether a comparable 'threshold problem' exists at the level of *jus in bello*. In other words, can radical asymmetry distort battlefield conditions to such a degree as to render uncertain what would otherwise be a morally uncontentious use of lethal force? At present, even when radical asymmetry is addressed in the context of *jus in bello*, analysis is mostly limited to its potential to jeopardize the safety of enemy non-combatants.

Risk-Transfer War

One of the major objections to casualty-averse Western warfare is that its mitigation of physical risk is too often obtained at the expense of the safety of enemy non-combatants. As with the 'threshold problem', OAF provides an early example of this critique. In November 1998, it was reported that NATO commander, Gen. Wesley Clark, had issued a 'no loss of aircraft' restriction. Though this was subsequently denied by Clark, he did write that 'nothing would hurt us more with public opinion than headlines that screamed, "NATO LOSES TEN AIRPLANES IN TWO DAYS"' (cited in Mason 2010: 239). In addition to ruling out the use of ground troops, NATO pilots were mostly restricted to high-altitude bombing. Flying at lower altitudes was discouraged even to deliver food supplies for thousands of displaced Kosovars, on the grounds that doing so would incur undue risk (Coker 2001a: 128).

[4] At its most extreme, this argument manifests as an overwrought concern that such circumstances may spell the end of democracy itself (Debrabander 2014). Another problem with much of this writing is its insufficient analysis of the subtle but crucial distinction between 'easier' and 'too easy' resorts to violence. In the case of humanitarian intervention, for example, many might welcome a lower political threshold for the use of force. See Beauchamp and Savulescu (2013) for more detail.

For many, NATO's commitment to force protection during the conflict did not so much eliminate physical risk as *transfer* it to the vulnerable non-combatants on the ground (Cook 2004: 127; Luttwak 1999: 41; Shaw 2005: 22; Shue 2003; Shue 2011: 137). The practice of high-altitude bombing, it was argued, generated excessive levels of collateral damage, and failed to prevent the ethnic cleansing on the ground.[5] OAF highlighted the extent to which casualty aversion had been internalized by the American military, with the avoidance of Allied loss of life functioning as the basic standard for judging military success (Record 2000: 5; Robinson 1999: 673).

In the years since, criticism of this type has re-emerged in response to the increasingly asymmetric nature of US warfare. Williams has argued that within Afghanistan, 'non-combatants will be protected so long as their protection does not require taking measures that may endanger the lives of soldiers' (2008: 589). Enemark makes a similar observation in regard to UAV violence, arguing that it 'constitutes the systemic transferal of risk from combatants to non-combatants' (2013a: 7). Radically asymmetric warfare is controversial to so many precisely because of its potential to continue and intensify this transferal. Any development in war that increases the already considerable burden of risk assumed by non-combatants warrants scrutiny. Problematically though, much of this analysis contains an implicit assumption that should non-combatant safety be restored, and killing restricted only to the liable, then the *in bello* challenge of radical asymmetry would be resolved. What this overlooks is the potential of this same asymmetry to destabilize the very moral frameworks responsible for this designation of liability. Determining the significance of this more subtle challenge requires analysis of the moral rules of inter-combatant violence, as well as the deeper assumptions of risk and threat that underpin them.

More Than a Question of Conformity

The Expendability of Combatants

Jus in bello has historically captured the moral rules that apply once hostilities have commenced. It is an evolving regime, subject to challenge, re-evaluation,

[5] In fact, the opposite is likely true. In the weeks preceding the conflict, then CIA Director, George J. Tenet, warned that Serb-led Yugoslav forces might respond to NATO's air operation by accelerating their campaign of ethnic cleansing (Kurtz 1999: C1). This prediction proved accurate and the Serb response to NATO high-altitude bombing was indeed an escalation of violence on the ground (Booth 2000: 8; Lambeth 2000: 196–7; Pevehouse and Goldstein 1999: 540; Spiegel and Salama 2000: 2206; Wheeler 2000: 269).

and revision. At the same time, however, it is anchored to a number of less contestable demands. Arguably the most significant of these today is the principle of discrimination. Non-combatants, it is held, are immune from direct and deliberate violence; they may suffer collateral harm only when the unintentional injury is proportional to the military objective desired.[6] In his formative work, *Just and Unjust Wars*, Michael Walzer divided the rules of war into two clusters of prohibitions; the first centred on when and how soldiers can kill, and the second, on whom they can kill (2006a: 41–2). According to Walzer, this second set of rules, grounded in non-combatant immunity, was fundamental to the very practice of war.[7] To follow this logic, when assessing the legitimacy of emerging weapons and techniques, including those that convey an asymmetric advantage to one side, it is to the potential fate of the non-combatant that we should first cast our moral judgement.[8]

Alongside the principle of discrimination exists the countervailing constraint of military necessity—combatants may be injured and killed, and military objects destroyed, in order to degrade the military capacity of the enemy. In accordance with this principle, the moral frameworks of war place few restrictions on the type and degree of violence that may be directed against opposing combatants.[9] Any argument that radical asymmetry constitutes a challenge to the moral right to kill must reconcile this claim with the highly permissive conditions of inter-combatant violence currently authorized by the moral frameworks of war.

Given the importance of non-combatant immunity, it should come as little surprise that within the literature itself, the *in bello* permissibility of asymmetric killing, including UAV violence, is most often determined on the basis of its

[6] The principle of proportionality functions within both the moral and legal frameworks of war to navigate the dual demands of discrimination and military necessity. According to the principle, though indirect civilian death is itself not a crime, attacks are prohibited in situations in which incidental civilian injuries 'would be clearly excessive in relation to the concrete and direct overall military advantage anticipated' (ICRC 1998: Art 8(2)(b)(iv)).

[7] See also, Johnson (1981; 1999: 37); and Ramsey (1968: 355).

[8] Non-combatant immunity is also regarded as one of the fundamental pillars of the contemporary laws of war. Spaight wrote in 1911 that 'the separation of armies and peaceful inhabitants into two distinct groups is perhaps the greatest triumph of international law' (1911: 37). The temporary collapse of this distinction in both world wars served as the impetus for the Fourth Geneva Convention, which established greater humanitarian protections for civilians in war zones (ICRC 1949). More recently, the International Court of Justice (ICJ) affirmed the importance of civilian immunity, describing it as a 'cardinal' and 'intransgressible' principle, which 'constitute[s] the fabric of humanitarian law' (1996, cited in Dill 2015b: 70). 'Remove the "principle of distinction"', argues Dinstein, 'and the entire IHL [international humanitarian law] system collapses' (2007: 146).

[9] Nor do the laws of war. Exceptions include a prohibition on 'weapons of a nature to cause superfluous injury or unnecessary suffering' (ICRC 1977). Exploding (ICRC 1868) and expanding bullets (ICRC 1899a), asphyxiating gases (ICRC 1899b), and anti-personnel blinding lasers (The Convention on Certain Conventional Weapons 1995: Art 1) have all been argued to fall within the scope of this ban. Other *in bello* restraints include prohibitions on perfidy, treacherous killing, and the threat to deny quarter (Goodman 2013: 827).

compliance with the principles of discrimination and proportionality (Ahmad 2013; Anderson 2009: 8; Barela and Plaw 2016; Bergen and Tiedemann 2010; Braun and Brunstetter 2013; Cortright and Fairhurst 2015: 7–8; Fricker, Plaw, and Williams 2010: 8; Kaag and Kreps 2014: 78–9; Kilcullen and Exum 2009; May 2013; Plaw 2010: 9; Zenko 2013a). This applies equally to literature dealing with the growing autonomy of robotic weaponry, a development that will likely enable even greater levels of military imbalance between conflicting parties (Arkin 2010; Canning 2008; Crootof 2019; Human Rights Watch 2012; Human Rights Watch 2013; Kellenberger and Spoerri 2011; Lin, Bejey, and Abney 2008: 76; O'Connell 2014; Sharkey 2008: 17; Sharkey 2010). Left comparatively unexplored is the potential of this same asymmetry to undermine the presumption of liability to direct and deliberate harm that would otherwise apply to the fighters of the weaker side.

Walzer writes that 'there is no principle of Just War theory' that prohibits the use of military force from distances sufficient to render the combatant 'largely invulnerable to counterattack... so long as they aim accurately at military targets' (2002: 936).[10] On the subject of UAVs, McMahan goes further, suggesting that the capability of this technology to be utilized in the absence of physical risk 'makes them unambiguously good in the hands of just combatants' (2013: x). According to a 2011 report by a British Ministry of Defence (MoD) think tank, the prevention of loss of aircrew life '*in itself* morally justifies the use of UAVs (UK MoD 2011: 56, my emphasis). Strawser argues that 'a technology that better protects (presumably) justified warriors is not merely morally permissible, but ethically obligatory' (2010: 343).

In light of the highly permissive rules of inter-combatant violence, these accounts of the moral status of asymmetric force are understandable. Nevertheless, they remain incomplete. Specifically, they fail to properly recognize that it is not merely through rule violation that the frameworks of war may be challenged. This was observed by Hedley Bull, who argued that in addition to non-compliance, there exist more subversive threats to the efficiency of rules:

> What is a clearer sign of the inefficacy of a set of rules is the case where there is not merely a lack of conformity as between actual and prescribed

[10] As Chapters 2 and 6 will highlight, Walzer has suggested elsewhere that there *does* exist a tension between asymmetric violence in war and the moral right to kill. He has expressed further concern that the threshold for authorizing UAV strikes will be lowered, on account of their mitigation of physical risk (Walzer 2013a).

> behaviour, but a failure to accept the validity or binding quality of the obligations themselves... [as well as]... the premises from which they derive.
> (1977: 133)

Too often, the issue of rule observance is addressed to the exclusion of a deeper analysis of the potential of radical asymmetry to destabilize the very premise upon which these moral rules rest. Norms, including those of war, operate at two levels. They first function instrumentally, proscribing or prescribing explicit behaviour. Norms also, however, operate more fundamentally, constituting the very identity and situation of actors (Farrell 2005: 1–8; Price 2007: 10; Wendt 1999: 92–138).[11] Constitutive norms give meaning to action by defining the structure in which such action is intelligible (Nyhamar 2000: 29). As already noted, combatants in war possess a moral licence to kill that far exceeds the scope of permissible violence in other settings. Nuremburg Trials prosecutor Telford Taylor argued that the state of war 'lays a blanket of immunity' over combatants, allowing them to engage in 'acts that would be criminal if performed in times of peace' (1970: 19). What are the assumptions that structure this permissiveness, and crucially, is there a point at which they may be so challenged as to cast doubt on the existing rules as a morally intelligible précis for appropriate conduct in battle?

A complete account of the challenge of radical asymmetry must consider the extent to which the coherence of the moral justifications for killing in war rests on *assumed* conditions of mutual threat and endangerment between antagonists. Obtaining this account requires an examination of the moral claims that underpin the existing rules, as well as the historical factors and debates that have contributed to their formation. This historical lens in particular has yet to be properly applied to this issue. A majority of those sources that have drawn upon history have done so superficially, to a degree that has served to obscure, rather than illuminate the moral significance of increasingly one-directional violence.

The Chivalric Dismissal

UAVs are not the first method of killing to attract criticism on account of their 'riskless' quality. Rather, there exists a *historical* unease towards

[11] For more information on the distinction between regulative and constitutive rules, see Rawls (1955: 24); and Searle (1995: 27).

emerging modes of warfare that severely distort the reciprocal risk of battle. Crucially though, in each of these historical cases, the violence in question has eventually normalized. This has led many authors to conclude that contemporary asymmetry presents no moral issue not already raised (and resolved) within prior episodes of military imbalance (Brooks 2012a; Gregory 2016; Hallgarth 2013: 27; Lamb 2013: 65). Strawser, citing the First Gulf War and Kosovo, typifies this position, writing '…if there is a moral problem here due to asymmetry, it seems to have occurred long before UAV implementation and is not endemic to them' (2010: 356). Brooks similarly suggests that any move to restrict the use of UAVs on the basis of their risk-asymmetry would be morally no different than stripping combatants of body armour and firearms or 'requiring troops to engage in hand-to-hand combat' (2012a).

This dismissal typically frames concern over radical asymmetry as merely the latest attempt to impose a chivalric standard of 'fair fighting' onto the battlefield. According to Coady, '[r]iskless warfare poses many problems, as does hegemonic military power, but the fair fight story is not one of them' (2008a: 175). The following quote, found in *Aerospace* magazine, further highlights this position:

> War is not a sport and there is no requirement for those involved to be at equal risk, let alone to expect one protagonist deliberately to increase the risk they face on the basis of some spurious notion of fairness. (2013: 16)

While these two statements are technically correct, they ultimately do little to advance our understanding of the type and degree of asymmetry we witness today. In relation to the moral justifications for killing in war, they leave unanswered the question of how critical a difference exists between the significant reduction of inter-combatant risk, observed throughout the history of war, and the complete or near-complete elimination of risk at the level of collective belligerents. To frame concern for unilateral battlefield violence in terms of 'equal risk' or a 'fair fight' is to mischaracterize it. The significance of this challenge stands not on whether warfare should be a *fair* fight, but rather if it should be a *fight* at all. Specifically, whether there exists an assumption of a minimum standard of mutual risk within the moral frameworks of war. Historical episodes of risk-disparity have much to teach us regarding the challenge and likely implications of radical asymmetry. But exactly how much can only be determined through *sustained* historical reflection, something that has yet to be undertaken within the literature. As the next section will address, this absence of historical analysis is equally conspicuous within those sources that do consider radical asymmetry an *in bello* challenge of significance.

Engaging the Challenge Itself

The first specific reference to the moral challenge of contemporary 'risk-free' warfare can be found in the work of Paul W. Kahn. Kahn viewed the high-altitude bombing conducted during the 1999 military intervention into Kosovo as a critical juncture in American warfare. At issue was the impunity with which such violence was applied (Kahn 1999: 201). These arguments were developed further by Kahn in *The Paradox of Riskless War*, which detailed the seriousness of the challenge to Just War principles represented by a US military policy able to actualize violence in the virtual absence of physical risk (2002). During this same period, Ignatieff voiced similar concern regarding the emergence of 'zero-casualty warfare'. Like Kahn, Ignatieff questioned whether the asymmetry imposed by American forces could be reconciled with the 'expectations that govern the morality of war' (2000: 161).

More recently, Kaempf has argued that the growing asymmetric potential of the United States military stands in tension with the moral justifications for lethal force (2014). This has been echoed by a number of authors, who view reciprocal risk as an essential condition of morally justified battlefield violence (Chamayou 2015; Enemark 2013a; Kels 2012; Williams 2015: 104; Zehfuss 2011). This analysis is overdue but nevertheless incomplete. According to Kaempf:

> If the fundamental principle of the morality of warfare that legitimises the killing of another soldier arises exclusively on the basis that such killing constitutes the right to exercise self-defense within the condition of a mutual imposition of risk, then the emergence of asymmetrical risk-free warfare represents a deep challenge. (2014: 79)

What Kaempf and others fail to properly convey, however, is just how large an 'if' this claim represents. Just War is not a settled theory, but rather a vast historical, heterogeneous tradition of philosophical and theological thought. It is a mutable framework, as we have seen most recently in relation to the revisionist critique of the conventional assumption of the 'moral equality' of combatants. Yet this complexity is largely overlooked within much of the literature that engages directly with the challenge of radical asymmetry. Rather than providing a systematic explanation of the role of reciprocal risk, too many authors merely take for granted its status as a fundamental condition in the licence to kill. Further analysis is needed of the precise function of reciprocity within each of the distinct and at times contradictory positions of the Just War Tradition. From this, the exact locus of the asymmetry-challenge can be established. Does this tension stem, for example, from the non-risk experienced by those who kill or from the non-threat projected by those targeted? Furthermore, what is the degree of risk that must be mitigated in order for this tension to emerge? Is it sufficient to impose risk-disparity between opposing *individuals* or is an elimination of risk required at the level of collective belligerents? What, in short, is the crossover point between unproblematic and problematic killing?

In determining this, it is again worth considering the distinction between the instrumental and constitutive function of norms. Just how foundational is the challenge presented by this intensifying military imbalance? This has already been examined within the literature by a number of scholars, who have argued that radical asymmetry challenges not only the justifications for killing, but also our very understanding of what forms of violence constitute war (Chamayou 2015: 91–2; Clark 2015: 6–7, 95–109; Enemark 2013a: 4, 59–62; Gross 2010: 268; Gusterson 2016: 147; Kahn 2002: 4). Just as with other literature, these accounts would be strengthened through greater historical perspective. How has war traditionally been understood and defined, and what role does reciprocal risk occupy within this understanding? Moreover, if these authors are correct and war has indeed changed in a fundamental sense, what are the implications of this, if any, for the contemporary regulation of lethal force? Existing sources struggle to answer these questions. This book expands upon this promising body of work to provide a comprehensive and historically informed account of the challenge and implications of radical asymmetry.

To summarize this section, much has been written on the growing ability of the United States to kill with impunity in war. Problematically, however, the full impact of this phenomenon has yet to be properly evaluated. Criticism of radical asymmetry, when it does emerge, is mostly limited to its tension with either democratic restraints on the use of force, or the immunity of enemy non-combatants. In contrast, its potential to challenge the coherence of the moral standards of inter-combatant violence has received little attention, particularly from a historical perspective. Too often, when history is drawn upon in this context it is used merely as a tool to pre-emptively dismiss the moral challenge of radical asymmetry. Typically absent from this dismissal is sufficient analysis of the precise degree to which current asymmetry reflects or diverges from the resolvable challenges of the past. This criticism also applies to those sources that have engaged more directly with this issue. Though a necessary addition to this debate, a majority of these sources have yet to provide a sufficient depth of analysis of the contemporary and historical role of reciprocal risk within the frameworks of war.

This chapter concludes by detailing how each of these gaps in the research will be addressed in the chapters ahead. Through a combination of historical study of the principle of reciprocal risk, and analysis of historical and contemporary episodes of asymmetric violence, this book will comprehensively address the moral significance and implications of radical asymmetry.

Historicizing the Challenge of Radical Asymmetry

Reciprocal Risk and History

History will serve a number of functions within this book. It will first be drawn upon to construct a plausible account of the role of reciprocal risk in the ethical and moral frameworks of war. The warrior ethos and Just War Tradition are important intellectual resources in the regulation of lethal force, both in their own capacity, and as sources of influence for the customary and positive laws of war. Recognition of this influence is especially important in relation to the warrior ethos, a framework that is too often overlooked when assessing the conduct of today's wars. The standards fostered within the warrior ethos, and in particular, the European rules of chivalry of the twelfth and thirteenth centuries, were a major historical basis for the laws of war (Solis 2010: 5). This influence is evident in the United States Department of Defense (DoD) Law of War Manual, which lists both 'honour' and 'chivalry' as vital components of the contemporary laws of war (2015: 66).[12]

Within both the warrior ethos and Just War Tradition, the justifications for inter-combatant violence will be examined, as well as the assumptions from

[12] This influence can also be found in the US Army Law of Land Warfare Manual, which requires its combatants to conduct hostilities 'with regard for the principles of humanity and chivalry' (US Army 1956: 3). The 1940 draft of these same rules 'denounces dishonorable means, expedients, or conduct' (US Army 1940: 2). The 1914 draft cites 'the principle of chivalry', which 'demands a certain amount of fairness in offense and defense and a certain mutual respect between opposing forces' (US Army 1914: 13). Further evidence of the influence of the warrior ethos can be found in the 1977 Additional Protocol I to the Geneva Conventions' (API) prohibition on 'perfidy' (ICRC 1977: Art 37). When justifying its inclusion, the *Commentary* on Article 37 made explicit reference to the principle of chivalry, stating that 'perfidy was considered a dishonour which could not be redeemed by any act, no matter how heroic' (Sandoz, Swinarski, and Zimmermann 1987: 434).

which they derive their coherence. This book will further establish the point at which reciprocal risk may be so undermined as to provoke a challenge to these assumptions. History will also be used to assess the implications of radical asymmetry for the regulation of lethal force. This book takes seriously the claim made by some that radical asymmetry represents a fundamental challenge to our shared conception of war. Particular focus will be given to the potential of this increasingly one-directional violence to incentivize a breakdown of battlefield restraint, among both the stronger and weaker party. Through its engagement with the warrior ethos and Just War Tradition, as well as war more broadly, this book will provide a theoretically and historically grounded account of reciprocal risk that both widens and deepens the conceptual terrain of radically asymmetric violence.

Any attempt to study war must be conscious of the way in which ideas of legitimacy have been constructed and reconstructed throughout history. The meaning of concepts such as 'combatant' and 'non-combatant', 'military necessity', and even 'war', has evolved in response to material and ideational changes in military practice, as well as prevailing moral attitudes, cultural norms, and societal expectations. It is for this reason that analysis of the challenge of radical asymmetry cannot be disembodied from the past. One of the central tasks of this book is to determine the extent to which—amidst the change and variance of the history of war—a thread of reciprocal risk has endured as an underpinning assumption in the justifications for lethal force.

The fidelity of this book is to an image of history as a contested realm of ideas, identities, norms, and values. This stands in contrast to the positivist historical accounts found most prominently in the work of IR 'grand theorists'. According to this 'laboratory' model of history, past events are constituted by a set of objective and immutable facts, which may be drawn upon to formulate general and empirically verifiable propositions regarding the nature of international relations (Reus-Smit 2008: 400). Within such approaches, the debates of the past often serve as little more than '*post hoc* legitimating devices' for the validation or discrediting of 'presentist' theories (Bell 2003: 154; Condren 2014: 38; Elman and Elman 2001: 7; Hobson 2002: 9). Waltzian neorealism and neoliberal institutionalism are especially susceptible to this, given that both structure their central theoretical claims around the 'transhistorical logic' of anarchy (Lawson 2012: 206).[13] What these approaches too often overlook, however, is the inherent difficulty in harnessing history's

[13] While susceptible, neither of these theories is *inherently* ahistorical. There is a considerable body of scholarship that, while drawing implicitly from neorealist sociology and rational choice theory, remains historical in its approach (Lawson and Hobson 2008: 421).

predictive capacity. As Johnson observes, '[t]he problem of historical generalisations...as predictors of the future is that, however large their temporal base, they are unable to take account of particular contingent factors in the society being studied' (1981: 12).

This should not be read as a dismissal of the utility of history in the study of international relations. Wight was correct when he described history as the *sine qua non* of international theory (1966). Bull agreed, listing history, along with philosophy and law, as essential components of the 'classical approach' to international relations (1966: 361). History has the capacity to aid in the formulation of theoretical arguments throughout the social sciences, and indeed, historical inquiry stands as a central methodological tool within this book. In order to be sufficiently warded against the charge of presentism, however, research must be conducted in a way that properly recognizes historical multiperspectivity. Scholars must accept the improbability of history providing inviolate and universal laws of international relations. Through the study of history one can, however, identify consistencies and reoccurrences, as well as discontinuities. 'History doesn't repeat itself', noted Mark Twain, 'but it often rhymes'.

In the spirit of this, historical research should be conducted in accordance with a set of more limited aims. Most importantly, historical knowledge should be judged not on the basis of its infallibility, but on its plausibility (Reus-Smit 2008: 405). This approach will inform the historical study undertaken in the chapters ahead.

While this book avoids any attempt at 'grand theory', it does seek to identify a meta-narrative of sorts with regard to the historical principle of reciprocity. In doing so, it also marks a break from 'radical historicism'. This mode of research is deeply contextualist, asserting that to read history 'via the construction of a broader narrative' is to wrongly manipulate it in service of particular political and ideological ends (Lawson and Hobson 2008: 427). Radical historicists would likely express deep scepticism at the idea of drawing upon history to construct an account of the moral significance and implications of radical asymmetry, a condition of battle that has until relatively recently been impossible to impose in any meaningful sense. Today's battles, they would note, are waged in a manner and against a societal backdrop that is fundamentally distinct from prior historical periods. Unavoidably, therefore, a majority of historical sources will offer little specific insight into a battlefield condition able to render armed and otherwise unimpaired enemy combatants physically non-threatening.

A full account of the challenge of radical asymmetry, however, must grapple with a range of factors, including the precise conditions by which individuals are rendered targetable on the battlefield. This determination of liability has been shaped considerably by the authors of the past—Just War scholars, jurists, military commentators, combatants, NGOs, states, as well as the international community more broadly. Though the asymmetric advantage possessed by the United States is itself unprecedented, the tensions, debates, and assumptions that structure the challenge of radical asymmetry are not. It will be the task of this book to highlight the role of reciprocal risk within the distinct, and at times conflicting frameworks of the warrior ethos and Just War Tradition. This analysis will also identify the way in which both frameworks have historically grappled with the emergence of asymmetric violence in war. Did this asymmetry strain the permissibility of killing in war, and if so, what was the locus of its tension—the lack of personal risk experienced by those who killed, or alternatively, the perceived absence of threat projected by those who died? In short, was it an agent or object-centric challenge?

Historical Cases

This historical study of reciprocal risk will be supplemented with a more specified analysis of two historical episodes and one contemporary episode of asymmetric violence. Through this approach, it will be established whether radical asymmetry is merely the latest in a long trend of resolvable tensions with battlefield ethics, or alternatively, a more fundamental break from the past.

Asymmetric Killing examines three episodes of battlefield asymmetry: military sniping, manned aerial bombing, and UAV-exclusive violence. These cases have been selected on the basis that each illuminates a key historical moment in which an emerging mode of warfare generated an imbalance of risk between opponents so profound as to call into question the permissibility of such violence. By studying the disruptive impact of these historical episodes, the challenge of radical asymmetry can be better understood. Equally, an exploration of the way in which these historical controversies were navigated and resolved can provide us with a clearer sense of the likely implications of radical asymmetry for the regulation of lethal force.

One consideration regarding the selection of historical cases was whether to include colonial violence between European and non-European powers. Excluding contemporary asymmetry, these military confrontations offer perhaps the clearest illustration of unilateral violence in war. Crucially though, as a consequence of cultural and racial animus, these imbalances of risk rarely provoked the same opposition that both sniping and manned aerial bombing did when directed against 'civilized' powers (Fuller 1923: 191–2; Hardinge, cited in Ellis 1986: 92; Reeves 1923, cited in Colby 1927: 279). For this reason, colonial violence, while explored briefly in the manned aerial bombing case, will not be a major focus in the chapters ahead.

In addition to the two historical cases, this book will examine contemporary UAV-exclusive violence—situations in which unmanned weaponry is used to the exclusion of a significant ground force presence. The issue of radical asymmetry goes beyond the UAV, encompassing a broader penetration of distancing and robotic technology into the landscape of war. At present, however, this mode of violence best illustrates the associated moral challenges.

Conclusion

When speaking in reference to the rapid proliferation of UAV technology, Singer drew a parallel with the aeroplane at the start of the First World War: 'at first it was unarmed and limited to a handful of countries...then it was armed and everywhere. That is the path we're on' (cited in Shane 2011). What was left unexplored in this same statement was the capability of this weaponry to impose a military imbalance between opponents without historical precedent. Problematically, the literature has yet to address the challenge and implications of this asymmetry with a focus that matches the intensity of its development. This is particularly evident in relation to the potential of radical asymmetry to challenge the moral justifications for inter-combatant violence. The extent of this challenge will be confirmed in the next chapter, which outlines the main contentions of this book.

2
The Moral Challenge of Radical Asymmetry

Introduction

'It's kill or be killed', observed one British soldier during the Second World War; 'if you think they won't do it to you, you're dead' (cited in Longden 2007: 30). Statements such as this, repeated in some form throughout the history of war, illustrate the clarifying effect of risk exposure in battle. For all the ambiguity that might surround the permissibility of armed conflict itself, the pervasive threat of the battlefield has long provided combatants with a powerful and uncomplicated rationale for their own violence—the prevention of injury and death, to both themselves and those that fight alongside them. The principal focus of this book is the extent to which this logic, and the moral frameworks of war more broadly, may be called into question when violence can be administered by one side with near impunity.

As Chapter 1 made clear, this issue has yet to be addressed by the existing literature with a sufficient depth of analysis. Too often, the tension between radical asymmetry and the moral right to kill is dismissed on the grounds that such imbalance fails to violate an explicit rule of war. What this dismissal largely overlooks is the extent to which the coherence of these moral frameworks rests upon an *assumption* of reciprocal risk between opponents. This lack of deeper focus is problematic, contributing to a range of claims regarding the significance and impact of radical asymmetry that are not yet sustainable. Even among those sources that have expressed concern over the moral status of this violence, insufficient historical and theoretical perspective has left commentators unable to articulate the precise scope and intensity of this challenge.

This book remedies this lack of reflection, providing a comprehensive account of the challenge and implications of radical asymmetry. The purpose of this chapter is to outline the theoretical foundations of this account and then systemize the main contentions that formulate it. The chapter argues

Asymmetric Killing: Risk Avoidance, Just War, and the Warrior Ethos. Neil C. Renic, Oxford University Press (2020).

DOI: 10.1093/oso/9780198851462.001.0001

that the coherence of the permissive moral standards of inter-combatant violence is grounded in an assumption of war as a site of reciprocal *structural* risk between belligerents. Radically asymmetric violence fundamentally challenges this assumption, and with it, the moral premise of the permissiveness itself. This loss of coherence undermines the capacity of the existing moral frameworks to effectively delineate the reach and limits of appropriate violence in battle. In addition, radical asymmetry erodes the distinction between war and judicial sanction to a degree that incentivizes a dehumanization of those targeted. Lastly, radical asymmetry narrows the scope of legitimate retaliation for the weaker side. In doing so, it makes more likely a displacement of violence away from the increasingly immune military targets of the strong and towards its vulnerable citizenry.

This chapter will first outline the long-standing centrality of 'violence' and 'contestation' within understandings of the *nature* of war. These features manifest as a structural reciprocity of risk between opposing belligerents. This same reciprocity, it is argued, functions as a key assumption in the moral justifications for killing in war. This first section concludes with an overview of the historical challenge of asymmetry. Having established its theoretical and historical framework, the chapter will present the central argument of this book. Radically asymmetric violence strains the very understanding of war as a site of reciprocal structural risk. In doing so, it challenges the coherence of the existing moral justifications for lethal force to a degree not achieved by prior episodes of military imbalance. The chapter concludes by clarifying the broader implications of radical asymmetry; including its potential to weaken essential restraints between warring parties.

The Structural Reciprocity of War

The primary focus of this book is the extent to which the moral justifications for battlefield killing are challenged by conditions of radical asymmetry. In order to establish this, however, it is first necessary to examine the conceptual category of violence within which such justifications are situated. There exists, argues Clark, a 'crucial interrelationship' between how war is conceptualized and the frameworks that develop to govern its conduct (2015: 23). The following section identifies structural reciprocal risk as a critical and enduring element of this conceptualization.

Violence and Contestation

Like all institutions, our understanding of war is embedded with a range of political and cultural assumptions. War was long thought of as instinctive, an unavoidable by-product of the genetic disposition of our species towards violence (Lorenz 2010). War has at other times been regarded as a divine judgement or legal trial, analogous to interpersonal morality and law. The emergence of the state brought with it a more instrumentalized conception of warfare, along with the subsequent relegation of non-state violence to criminality (Bull 1977: 268–70; Devetak 2008: 11; Fullinwider 2003: 23; Johnson 2011: 9; Rousseau 1968: 56). In the mid-nineteenth century, 'total war' emerged—the military mobilization of entire societies, combined with the commitment of participants to unrestricted violence. As this variance should clarify, the designation of war is governed to a significant degree by a social and political calculus (Clark 2013: 37–61), as well as prevailing moral and legal convention (Finlay 2010: 312). The extent of this variance necessarily complicates the search for a historically fixed definition of war.

This is not to argue that all aspects of war are in perpetual flux. Across the historical record, one can also locate a set of deeper assumptions regarding the *nature* of war—a basic material structure that endures regardless of shifts in its technological, political, cultural, or moral *character*. This chapter argues that both 'violence' and 'contestation' are fundamental components of this presumptive nature. This violent contestation manifests as structural reciprocal risk—a sustained and pervasive imposition of danger upon and between belligerent parties. These features do not encompass all that there is of war, but they do delineate it in an essential sense. Scarry outlined this in her influential 1985 book, *The Body in Pain*. According to Scarry, war contains two constitutive features:

> [F]irst, that the immediate activity is injuring; second, that the immediate activity of war is a contest. In participating in war, one participates not simply in an act of injuring, but in the activity of reciprocal injuring where the goal is to out-injure the opponent. (1985: 63)

War has many variants. It may be inter-state or intra-state; religious or secular; limited or total; humanitarian or opportunistic. Crucially though, each of these models has traditionally been grounded in an assumption of structural reciprocal violence.

The centrality of violence in the nature of war should come as little surprise. As Scarry herself notes, violence 'is war's product and its cost, it is the goal toward which all activity is directed and the road to the goal' (1985: 80–1). This basic fact has been observed throughout history, including within the works of Prussian general and military theorist Carl von Clausewitz. In his classic text, *On War*, Clausewitz left no illusions as to his understanding of the role of violence in war:

> Kind-hearted people might think that there was some ingenious way to disarm or defeat an enemy without too much bloodshed and might imagine this is the true goal of the art of war. Pleasant as it sounds; this is a fallacy that must be exposed. (2008: 13)

Intense and collective physical violence has long been regarded as an essential characteristic of the nature of war. Fidelity to this conception has led scholars such as Gray (2001: 61), Coker (2014), and May (2015a) to exclude a range of non-kinetic, 'bloodless' forms of conflict, including hostilities in cyberspace, from the category of war. 'War', Gray argues, 'is about killing and maiming human bodies, and if that does not happen, it is not war' (2001: 61).

War is perceived to hinge on violence, and specifically, collective and *reciprocal* violence. This contestation dynamic has long distinguished war from massacre, genocide, and other instances of large-scale, one-directional violence, as well as collective policing. Cicero famously described war as a 'contending' by force, referencing the violence applied and absorbed by each party (cited in Grotius 2012: 23). Centuries later, Dutch jurist, Cornelis van Bijnkershoek wrote that 'war…is a contest' (1810: 1). 'A contest of force' was also the description advanced by American jurist and diplomat Henry Wheaton (1836: 212). Military theorist Francis Lieber subscribed to this as well, defining war as a 'protracted contest' between 'two considerable masses' (1889: 630).

To describe the nature of war as one of contestation is not to diminish the seriousness of it in any way. Contest in these terms is not to be understood as synonymous with 'game'. The sheer discrepancy in the scale of the consequences is enough to render comparisons between war and gaming obscene (Scarry 1985: 83). What this term does highlight, however, is the assumed reciprocal dynamic at the centre of war. Threat, violence, and risk are two-directional, simultaneously, though not equally, projected and received by both parties. This was outlined by German jurist, Samuel von Pufendorf, who wrote of war as the 'state of men mutually [engaged] in offering and repelling injuries' (1729: 6).

In few works is the idea of contestation, and by extension reciprocity, so clearly detailed than within Clausewitz's *On War*. Clausewitz was dismissive of 'pedantic, literary definitions of war', opting instead to 'go straight to the heart of the matter' (2008: 13). War, according to Clausewitz, was:

> [N]othing but a duel on a larger scale... a picture [of which] can be formed by imagining a pair of wrestlers. Each [trying] through physical force to compel the other to do his will... Force to counter opposing force... (2008: 13)

Implicit within this image is a notion of war as a contestation of violence, as well as risk. For Clausewitz, defencelessness was the state to be inflicted upon the enemy, not the starting point from whence to commence attack (2008: 15). This interpretation is shared by Enemark, who views a reciprocity of risk as 'elemental' to the Clausewitzian model of war (2013a: 60).

Reciprocal risk is a fundamental and *enduring* assumption in the nature of war, as evidenced by more recent accounts. In 2011, Strachan and Scheipers provided the following criteria for war:

> [W]ar rests on contention. If one party attacks another, the other must respond for war to occur, or else what will follow will be murder, massacre, or occupation. This reaction means that possibly the most important feature of war is reciprocity. (2011: 7–8)

Strachan provides a similar definition elsewhere, describing reciprocity as a 'key element' in the nature of war (2010: 6).[1] It is important not to confuse an assumption of structural reciprocity with an expectation of true symmetry.[2] This is necessary to reiterate, for as Chapter 1 observed, concern regarding the moral challenge of radical asymmetry is too often dismissed as an anachronistic desire to transform war into some kind of ritualized pageantry of perfect equality. Belligerents seek to maximize their own military advantages—through strategy, tactics, training, and technology—while struggling equally to neutralize those of the enemy. Crucially though, this struggle to impose military overmatch is just as much a motivating factor for one's adversary as

[1] See also, van Creveld (1991: 159).

[2] Anything close to a structural *equality* of risk between opponents has been understandably rare throughout the history of war. However, approximations have emerged in response to a number of factors, including uniformity in tactics and troop composition (warfare between ancient Greek Hoplite Phalanxes); analogous military and economic strength (Iran–Iraq War—1980–8); as well as shared geographical and technological constraints (trench warfare on the Western Front—1914–18).

oneself. This, in combination with inescapable technological and geographical limits, has for most of history been sufficient to ensure a 'rough equality of capabilities' between adversaries (Ryan 2012: 155–6). Russian Gen. Mikhail Dragomirov observed this in 1890, when writing of the duty of the soldier:

> We should not forget that our mission is to kill and be killed. We should never close our eyes to that fact. Making war by killing without being killed is a chimera; making war by being killed without killing is inept. So one must know how to kill, while being ready to die oneself...'
>
> (cited in Chamayou 2015: 97)

War is a contest, albeit one that is deadly serious, and frequently unequal. Alongside violence, this represents a fundamental assumption regarding the nature of war, one that is 'elementary and constant regardless of other contingencies of form' (Scarry 1985: 87). This contestation of violence translates into a structural reciprocity of risk between belligerents. This same reciprocity, it is argued, plays a key role in the moral justifications for lethal force.

Justified Killing in War

The Just War Tradition and warrior ethos both function to regulate the violence of combatants. The following section will examine the first of these frameworks. It argues that structural reciprocal risk operates as a core assumption in the permissive moral standards of inter-combatant violence.

One of the principal tasks of the moral framework of *jus in bello* is to prohibit intentional—and mitigate unintentional—violence against non-combatants. Outside this strict constraint, combatants are afforded considerable licence in war. The principle of military necessity permits soldiers to kill their enemy counterparts in virtually all circumstances (Walzer 2006a: 138). This permissiveness is tempered only somewhat by the principle of humanity, which prohibits the use of means and methods of warfare that cause 'superfluous injury or unnecessary suffering' (ICRC 1977), as well as acts such as the denial of quarter. Crucially, within none of these rules is the moral permissibility of inter-combatant violence made contingent upon a reciprocal imposition of physical risk between individual enemy combatants. Immediate and personal vulnerability to harm is not a moral precondition for justified killing in war.

It is argued here, however, that the *coherence* of these rules and standards is ultimately grounded in, and contingent upon, an assumption of war as a site of reciprocal *structural* risk. To be clear, this refers to inter-belligerent, rather than inter-combatant reciprocity. Structural reciprocal risk, while not a formal requirement itself, provides the tacit rationale for the permissiveness of the moral standards of war.

Recall that war is essentially a two-level exercise. It functions at once at the individual level: as a violent interaction between combatants; and at the collective, or structural level: as a violent contestation between belligerent forces. When adversaries confront one another on the battlefield, they do so primarily as representatives of their respective collectives (Dill and Shue 2012: 313; Rodin 2002: 6).[3]

The combatant, in turn, functions as a two-level threat. He or she is presumed to represent both a specific individual threat, and a contributing element of the overall collective threat of their side (Dill and Shue 2012; Kershnar 2005: 130–1; Rodin 2008: 48; Walzer 2006b: 4; Zohar 1993: 615).[4] It is this presumption—that collective belligerent forces labour under mutual conditions of threat, and by extension risk—that forms the basis for deontological justifications for the lethal targeting of their constitutive parts; i.e. combatants. The principle of self-defence is fundamental to the moral exculpation of killing in war. Crucially, though, it is an attenuated conception of self-defence, linked to the structural, rather than individual reciprocity of risk between contesting foes.[5]

Structural reciprocity has an equally significant role in consequentialist justifications for killing in war. The contractarian account of Just War frames non-combatant immunity as the outcome of an implicit bargain between adversaries, grounded in the shared understanding that combatants on both sides are liable to direct targeting in their stead (Benbaji 2008: 487; Lazar 2015: 126–7; Statman 2012: 97, 106). It is argued that the coherence of this division between protected and targetable rests upon an assumption of structural reciprocal risk between belligerents.

[3] Rousseau drew attention to this in his description of war as fundamentally 'not a relation between men, but between states…' (1968: 56). Though the participants of war today extend beyond the state, 'the collective' endures as the primary level of analysis.

[4] Lazar describes this two-level threat as a combination of 'micro-threat' and 'macro-threat' (2010: 186).

[5] Chapter 4 explores this in detail in the context of both conventionalist and revisionist accounts of Just War. It argues that even within the revisionist position, which rejects the idea of collectivized liability, an assumption of structural reciprocal risk underpins the permissiveness of its moral standards of violence.

Before concluding this section, it is important to reiterate the role of reciprocal risk within the moral frameworks of war. The coherence of these justifications for the use of lethal force does not rest, as some authors have claimed, on the maintenance of reciprocity at the level of individual combatants. Kahn is among those who suggest as much, arguing that 'combatants are allowed to injure each other just as long as they stand in a relationship of mutual risk' (2002: 3). Walzer seemed to echo this sentiment when, citing Camus, he wrote: 'you can't kill unless you are prepared to die' (2004a: 101). Both these claims are mistaken, no matter the Just War position one adopts. An assumption of mutual vulnerability is indeed a necessary condition for justified violence in war, but crucially, vulnerability at the level of collective belligerents, not the individual combatants themselves.

By this logic, military imbalance, if a challenge at all, should only present itself as such when imposed at the structural level of war. This contention, however, is complicated by history. Beginning with the most basic forms of ranged killing, modes of warfare have emerged that have mitigated the reciprocal risk between opposing *individuals* in battle to such a degree as to call into question the legitimacy of such violence. Is this not sufficient evidence, one might ask, of the necessity of individual risk in the moral justifications for killing in war? In order to illustrate why this is not the case, a more detailed overview of these historical challenges is needed.

The Historical Challenge of Asymmetry

In determining the moral significance of radically asymmetric violence, history is an indispensable source. Of particular value is the range of historical episodes in which a disparity of risk between opponents generated a comparable tension. This book examines two of these historical cases—military sniping and manned aerial bombing.

From their emergence, one of the most notable features of both sniping and manned aerial bombing was their ability to physically dislocate the operator from the immediate threat of retaliatory harm. This distancing of force generated a profound imbalance of risk between the combatant and target, at least in comparison to military alternatives of the period. It also provoked opposition from those that regarded such advantage as a battlefield transgression. In the case of military sniping, evidence suggests that from its earliest use the practice was widely criticized as dishonourable (Deforest 1867; Earley 2014: 182; Faust 2009: 42; Furgurson 2001: 173; Moore 1862: 73;

Pindell 1993: 46; J. Taylor 2011: 93; Wilkeson 1886: 120–1). So resented were its practitioners that upon capture snipers were often executed, rather than extended prisoner of war rights (Earley 2014: 182; Faust 2009: 41–2; Katcher 2002: 7; Mortier, cited in Thornton 1806: 2:59). Perhaps more tellingly, opposition to sniping can also be found among those who benefited from the practice—snipers themselves and those who fought alongside them (Crozier 1937: 101; Homer, cited in Johns 2002: 35; Ray 2006; Wilkeson 1886: 120–1).

A similar degree of opprobrium emerged in relation to aerial bombing, beginning with military ballooning (Christopher 1956: 27) and intensifying with the development of the manned bomber aircraft. For many, the bombing of ground-based targets was an ignoble act, particularly when compared to the highly venerated practice of aerial fighting (Allen 1972: ix; Crozier 1937: 104; Hastings 2013: 459; Lewis, cited in Robinson 2006: 155; Meijering 1987: 23; Molter 1918: 121; Powell, cited in Lance 1976: 67). The critical component in this distinction was the perceived lack of inter-combatant risk that characterized the former. In both cases—military sniping and manned aerial bombing—the opposition that emerged in response to this asymmetry is best understood in reference to the warrior ethos.

The warrior ethos has historically operated as a system of values and beliefs, created and internalized by fighters as a means of safeguarding their own dignity and self-respect. The ethos itself is comprised of a set of virtues, including courage, honour, loyalty, integrity, and selflessness. Within this ethical framework, the exposure of combatants to personal, physical risk has long represented an essential element in the maintenance of their privileged status. This is illustrated most starkly in ancient and medieval iterations of this battlefield code. It is argued that across these periods, the ethical legitimacy of violence has hinged to a significant degree on an expectation of inter-combatant risk. Honour could be accrued on the battlefield only when warriors directed their violence against those with the capacity to threaten them (Cicero 1850: 151; Coker 2002b: 17; Hanson 1989: 14; Tyrtaeus, cited in Lattimore 1955: 14; Meagher 2014: 28; Pizan 1999: 171; Wheeler 1990: 124).

This highlights an important distinction between the core motivation of the warrior ethos and that of the Just War Tradition. The warrior ethos has long preconditioned legitimate violence on the expression of virtues such as physical courage. Reciprocal risk between opponents provided the means of expressing such courage and was regarded, therefore, as a desired end in itself. This conception is agent-centric; the combatant's exposure to harm serves as the principal basis for his or her ethical right to kill. Without risk, the violence of the combatant is stripped of its extenuating heroism (Coker 2001a: 37). The Just War Tradition, in contrast, is primarily concerned with the inherent rights of the potential *targets* of violence. In this context, reciprocal risk is a side effect of the main objective of restricting violence to the materially threatening. The asymmetry-challenge generated by both military sniping and manned aerial bombing was primarily grounded in the non-risk to the user, rather than the overall non-threat of the target. What emerged in response to this perceived 'non-risk' was a contested narrative of legitimacy.

In order to account for the resolution of these historical challenges, it is first important to recognize that Walzer was essentially correct in describing the rules of war as 'a social creation' (2006a: 43). The norms of war are unfixed; 'what is and is not considered acceptable behaviour in war is historically determined, neither self-evident nor unalterable' (van Creveld, cited in Parker 1994: 40).[6] This is equally true of the warrior ethos itself, which is intended to reflect, and to a considerable degree, epitomize, the society within which it operates. It is an institutional expression of our values and must, therefore, by necessity, 'accord with the prevailing cultural climate...' (van Creveld 1991: 89). Inevitably, as the character of war—as well as that of society more broadly—evolves and modernizes, accepted understandings of honour and martial virtue will shift accordingly.

In relation to the challenge of asymmetry, some of the more significant of these changes to the character of war have been the progressive reduction in combatant agency (Blum 2010: 76; Constant 1988: 55; Jünger 1929: 180; Strachan 2014: 49), as well as a greater emphasis on principles such as restraint and professionalism in the ethical status of fighters (Freire 2010; Kirkpatrick 2015: 210). As the warrior ethos has evolved to better reflect these changes, as well as the ongoing demands of military necessity, physical courage and direct risk exposure have increasingly been regarded as a necessary, but insufficient, determinant of the ethical legitimacy of combatants. A key component in the gradual ethical normalization of both military sniping and manned aerial bombing was their potential to be conducted in accordance with this more contemporary ethos (Coughlin, cited in Wood 2015; Nelson 1992: 21; Shaw 2005: 21; Thomas 2001: 88; Wald, cited in Ignatieff 2000: 101).

Given the historical consistency of this resolution, one could be forgiven for assuming that a similar trend will apply to radical asymmetry. Indeed, it is argued within much of the literature that in terms of their asymmetric

[6] To be more specific, norms in the sense of perceived appropriateness or legitimacy are unfixed. Those represented by treaty law are by their nature far less subject to revision.

potential, UAVs fail to present any new problems not already generated by previous methods of 'killing from a distance' (Brooks 2012a). Whatever the appeal of this logic, categorical statements of this type are in error. Crucially, they fail to observe the critical distinction between an erosion of reciprocity between individual combatants and its mitigation at the structural level of war.

The Challenge of Radical Asymmetry

Radical asymmetry—manifested most starkly through the UAV-exclusive violence of the United States—represents a fundamental disjuncture from previous challenges of this type. Both military sniping and manned aerial bombing exacerbated the risk imbalance between operator and target to a degree that challenged the ethical status of the former. Neither, however, fundamentally undermined the *structural* reciprocity of risk that binds belligerent forces at the collective level. It remained possible in both historical cases to maintain the traditional, and necessary, presumption of the targeted enemy combatant—as an empowered representative of a significant collective threat. Radical asymmetry, it is argued, is distinct, eroding structural reciprocity to a degree that calls into question the very coherence of the permissive moral standards of inter-combatant violence.

The historical cases do in fact offer an illustrative precursor to this more significant challenge, in the form of contemporary manned aerial bombing. During both the 'Highway of Death' episode of the First Gulf War, and the high-altitude bombing of OAF, the military imbalance between the United States and its enemies drew widespread criticism. Just as with earlier cases, this opposition was partly grounded in the perceived tension between the impunity of American violence and the warrior ethos. At the same time, there were some who went further, questioning whether the presumption of moral liability among the combatants of the weaker side could be maintained in such conditions (Ignatieff 2000: 161; Kahn 1999: 201; Lazar 2015: 117; May 2015b: 93; Sanchez and Phillips 2008: 80; Waller and Barry 1992: 16). Crucially though, even within the context of OAF, although the United States achieved its objective of zero combat casualties, this outcome was not guaranteed. The physical presence of American pilots, even from the altitude at which they operated, necessarily imposed some degree of structural risk, i.e. a symmetrical potential to die. Though driven close to its vanishing point, reciprocity endured.

The substitution of a human fighting force for machine proxies, as has occurred in theatres such as Somalia, Pakistan, and Yemen, is distinct, concluding the long-standing configuration of the battlefield as a realm of mutual structural risk. What is imposed in its stead is the closest approximation yet achieved to perfect military asymmetry—a unilateral projection of violence by weaponized aerial robots. In many cases, targeted enemies have not only been rendered non-threatening at the precise moment of their incapacitation or death, but have been stripped—on aggregate—of their very *capability* to direct lethal force against the military personnel of the opposing side. This erosion of structural reciprocity represents a profound moral crossover point, in which the violence of American forces has become so one-directional as to undermine the moral basis for its permissiveness.

This is not to argue that UAV technology is intrinsically unjust. What is significant here is the resulting imbalance of risk and threat, not the precise means by which such imbalance is imposed. It is simple enough to imagine future conditions in which alternative technologies generate comparable or even greater degrees of risk-asymmetry. It is also important to acknowledge that the use of UAVs in 'hot' war zones, such as Afghanistan, where they are deployed alongside American 'boots on the ground', does not rise to the level of *structural* asymmetry necessary to constitute a challenge to existing *in bello* justifications for lethal force. When deployed to the exclusion of a significant ground presence, however, UAV violence generates a profound normative challenge.

First, as with previous episodes, radical asymmetry stands in tension with the warrior ethos. Targeted UAV strikes have unsurprisingly been characterized as cowardly and dishonourable by those subjected to them (Blair 2011: A21; Exum, cited in Mayer 2009). Perhaps more interestingly, this practice has also drawn criticism from those that fight alongside UAV operators (Burridge, cited in Mayer 2009; Chapelle, cited in Woods 2015: 179–80; Power 2013). Though this ethical challenge is significant, just as with sniping and manned aerial bombing, it is likely resolvable. Evidence already exists of counter-narratives emerging, in which the violence of UAV operators is legitimized through an appeal to professionalism and restraint, rather than physical courage (Mathewson, cited in Enemark 2013a: 86; Lee 2014: 20). It is here, however, that the comparison between historical and contemporary asymmetry concludes. UAV-exclusive violence, through its strain on structural reciprocal risk, goes beyond a mere tension with the warrior ethos, challenging the coherence of the existing moral justifications for killing in war.

One potential response to this claim is a rejection of the empirical basis of radical asymmetry. War possesses an interactive dynamic, within which, the duel between opponents can take different forms. Those subjected to military imbalance are not passive objects, but rather active participants who will seek to contest an unfavourable status quo. This may include a redirection of violence by the weaker party to the more vulnerable civilian targets of the strong. The pervasiveness of terrorism today provides a clear illustration of this dynamic. Though UAV-exclusive violence has succeeded in insulating the military forces of the United States from reciprocal harm, the American public remains under the general threat of retaliation in the form of terrorist attacks. In this sense, it is perhaps more accurate to regard the physical risk sustained by the United States as *diffused*, rather than transcended. This aligns with Walzer's description of terrorism as the production of 'general vulnerability' (2004c: 51–2). Even if we accept this, however, we are left with a fundamentally altered form of threat, insufficient to morally justify the current permissiveness of American violence. This can be observed through a more detailed analysis of the Just War positions of conventionalism, contractarianism, and revisionism.

Conventionalism has long represented the orthodox moral position within Just War. Arguably its fundamental claim, promoted most famously by Walzer, is the 'moral equality of combatants'. According to this doctrine, fighters on both sides, regardless of the *ad bellum* status of the conflict, possess identical privileges and responsibilities, including an equal right to kill. Violence is subsequently justified according to the principle of self-defence. The physical threat projected by opposing combatants renders them mutually liable to retaliatory harm (Enemark 2013a: 81; Kenny 1985: 10; Nagel 1972: 128; Norman 1995: 168; Ryan 2008: 138; Shue 2004: 111; Shue 2008: 100; Walzer 2006a: 145; Zehfuss 2011: 555). Non-combatants, in contrast, are presumed to be materially non-threatening, and on those grounds, protected from direct and deliberate violence. As already highlighted though, this position assumes a collectivized understanding of liability. Through their military membership, combatants are judged to sufficiently contribute to the 'unified threat' of their side, and may, on this basis, be targeted for injury and death (Walzer 2006b: 4). It is argued that radical asymmetry fundamentally undermines this designation of 'unified threat'. When the collective threat of the enemy has been reduced from a sustained and pervasive military danger to something approximating sporadic crime, the moral force behind 'collectivized' targeting is weakened.

The moral challenge of radical asymmetry can also be witnessed in the context of contractarian accounts of Just War. According to this position, the permissiveness of inter-combatant violence, and inverse prohibition against non-combatant targeting, is the outcome of a rational bargaining process. Both sides accept this division on violence because it offers a mutually beneficial way to administer and limit the destructiveness of war (Benbaji 2008: 487; Brandt 1972: 149–50; Schelling 1980: 6). It is argued that the imposition of radically asymmetric violence amounts to a tacit withdrawal by the stronger party from this implicit bargain. Absent structural reciprocity, the coherence of this moral justification for killing in war is undermined.

The moral tension of radically asymmetric violence is less direct within the revisionist position of Just War. This relatively recent challenge to the conventionalist account makes no distinction between the moral landscape of war and that of normal society, applying identical standards of liability to each. The first implication of this is a rejection of the moral equality of opposing combatants. Rather, revisionist scholars precondition the right to kill in self-defence on a test of just cause (Fabre 2014; Frowe 2014; McMahan 2008: 22; Rodin 2002; Strawser 2011). The basis for moral liability shifts from material threat, to moral responsibility for a material threat. The second implication of this conception is a rejection of collectivized designations of liability. The targetable status of enemies is instead calculated on an individualist basis (McMahan 2009: 156). At first glance, radically asymmetric violence would appear to present no challenge at all to this position, at least for as long as it is directed by the just against the unjust. It is argued, however, that even within revisionism, radical asymmetry has the potential to test the moral right to kill, by undermining the *presumption* of liability among the weaker party. This challenge is most evident when those targeted on the unjust side are low-level enemies.

Recall that in contrast to their political leaders, military commanders, as well as certain influential civilians, the overwhelming majority of ordinary combatants will make no meaningful political or military contribution to the overall course or conduct of the war in which they participate. Instead, their contribution to the 'business of war' is almost entirely contingent on the physical threat they represent to the enemy. Within conditions of radical asymmetry, this threat may be so reduced as to push their contribution below the level sufficient for moral liability to harm. Grappling with the implications of this is essential, particularly given the frequency with which UAV-exclusive violence has been directed not against mid- and high-level targets, but rather low-level fighters (Barrinha and Vinha 2015: 23; Bergen and Rowland 2012;

Brooks 2013; Brooks 2014; Devereaux 2015; Entous 2010; Human Rights Clinic 2012: 8; Landay 2013; Linden 2015:173; Telhami 2004: 16; Woods 2015: 20). In these cases of low-level targeting, the coherence of a presumption of liability founders.

This is not to suggest that low-level fighters of the weaker side are no longer targetable. This book does not (and could not plausibly) argue that once asymmetry between opponents reaches a certain level, the violence of the stronger side is automatically rendered impermissible. What it does argue is that within these conditions it is significantly more difficult to maintain the *presumption* of liability to harm.[7] Recall that civilians themselves also contribute to the collective threat of their side in a time of war. They feed hungry armies, elect bellicose leaders, and educate future combatants. They maintain their protected status, however, on the basis that this contribution is generally insufficient to invalidate their immunity from direct and deliberate harm. One of the most significant challenges of modern counter-insurgency warfare is identifying enemy fighters and then directing violence towards them and away from non-combatants. Radical asymmetry threatens to exacerbate this challenge by obscuring the very basis of moral differentiation between these two groups.

The permissiveness of inter-combatant violence rests upon an assumption of reciprocal structural risk. Radically asymmetric violence undermines this assumption, and with it, the coherence of the existing moral frameworks of war. In the absence of this coherence, the status of these frameworks as an effective guide for appropriate conduct in battle is strained.

There is an important qualifier to this argument. As outlined above, the key element in the moral challenge of radical asymmetry is the extent to which it invalidates the *threat* of the weaker party. It is possible, therefore, to imagine conflict conditions similar to a law enforcement scenario, in which military forces apply radically asymmetric violence to a target group, who retain their liability to harm on account of the physical threat they pose to a third party. As Christian Enemark has observed, 'the unidirectional application of force might be consistent with acting justly if the purpose is law enforcement' (2013a).[8] Chapter 6 will explore this in greater detail, in the context of the high-altitude bombing of OAF.

[7] Even if the conduct of these low-level individuals falls short of liability for summary execution, they may remain, in many cases, justifiable targets for some degree of moral and legal sanction.

[8] See also, Kahn (2002).

Before this, however, the significance of this shift in the logic of hostilities must be properly recognized. Chapter 1 noted that for some within the literature, the challenge of radical asymmetry extends to our very understanding of what forms of violence constitute war. The next section examines this claim, along with its implications, arguing that radical asymmetry has the potential to trigger a more general breakdown of battlefield limits. First, by promoting the dehumanization of the weaker side; and second, by incentivizing those targeted to intensify attacks against the non-combatants of the strong.

The Changing Nature of War

In his influential book *The Utility of Force*, former British officer Rupert Smith explored the level of change undergone by war throughout the twentieth century. He followed this analysis with a deliberately provocative claim, arguing:

> War no longer exists. Confrontation, conflict and combat undoubtedly exist all round the world...None the less, war as cognitively known to most non-combatants, war as battle in a field between men and machinery, war as a massive deciding event in a dispute in international affairs: such war no longer exists. (2005: 1)

What Smith referenced here was the observable decline in warfare as a confrontation between two comparable standing armies on a clearly delineated battlefield. Yet what claims of this type fail to properly credit is the extent to which the character of war is always evolving, in response to a range of internal (technology, military doctrine, etc.) and external (cultural, social, and political) triggers. The historical consistency and intensity of this adaption should give us pause before labelling any single military change, no matter how profound, as the 'end' of war.

The obvious question this raises is under what circumstances, if any, could we confidently declare otherwise; that yes, the material conditions of war *have* shifted to a degree sufficient to bring about a state of 'non-war'? Such a transition would be contingent upon the enemy no longer functioning as a responsive force, either by choice or as a consequence of insufficient capability. The radical asymmetry we are witnessing today, though historically unprecedented, does not yet achieve this. As already acknowledged, even within the context of UAV-exclusive violence, those targeted retain some capacity, albeit severely limited, to assert their will. Denied accessible military targets, this includes a redirection of violence towards civilians in the West.

Retaining radical asymmetry within the category of war does not diminish its significance. There are material constancies to warfare that if pressured, signify a change not merely in its character, but in its very *nature*. Structural reciprocal risk is one such constant. While radical asymmetry does not mark the *end* of war, it does challenge a set of deeper assumptions regarding its nature.

What then are the implications of this change in the nature of war for the regulation of lethal force? This section argues that beyond its potential to render less tenable the presumptions by which moral liability to harm is designated, radical asymmetry marks a significant shift in the basic logic of hostilities, from an adversarial and two-directional contestation, to something approximating judicial sanction. One outcome of this change is a greater likelihood of the dehumanization of the weaker party.

A Gardener's Vision of War

Before detailing the problematic implications of radical asymmetry, it is important to note that within classic Western Just War accounts, the image of war as a collective judicial exercise actually pre-dates the war-as-duel model outlined in the opening section of this chapter. Among the first to articulate this judicial account was the fifth-century philosopher and theologian Saint Augustine of Hippo.[9] The image of war as legal sanction can also be located in the works of Thomas Aquinas (Lackey 2008: 414; Reichberg 2008: 196), Francisco de Vitoria, and Francisco Suárez.[10]

Crucially though, whatever the reflections of Augustine, Aquinas, and others during this period regarding the *telos* of war, its material condition—its *nature*—remained one of contestation, grounded in a structural reciprocity of risk between opponents. Within conditions of radical asymmetry this

[9] Augustine's principal contribution to the Just War canon was the distinction he drew between the legitimate act of killing in war and the illegitimate act of murder. Importantly though, this allowance applied only to those combatants who employed such violence in response to a violated right: 'it is the iniquity on the part of the adversary that forces a just war upon a wise man' (Augustine, cited in Reichberg 2008: 195). Augustine framed the moral relationship between warring parties in strictly asymmetrical terms: 'one as an offender, the other as a dispenser of justice' (Haggenmacher 1992: 435). This was re-emphasized in Gratian's twelfth-century work, *Decretum*, which cited Augustine: 'Just Wars are usually defined as those which have for their end the avenging of injuries' (cited in Reichberg 2013: 157).

[10] In contrast to Augustine and Aquinas, Vitoria and Suárez conceptualized Just War more in terms of 'restitution' than 'punishment' (Reichberg 2008: 197–9). The dominance of this conception began to erode towards the end of the Middle Ages (Fulgosius 2006: 229), and particularly from the eighteenth century onwards, as evidenced by the work of Christian von Wolff (1679–1754), and Swiss jurist Emmerich de Vattel (1714–67) (Kalmanovitz 2015; Reichberg 2008: 193).

nature has shifted to allow a monopolization by the stronger party of both moral authority and violence. In a 1999 article on the Kosovo bombing, Kahn described the logic of the United States thusly: 'if our end is virtuous, there can be no justification for suffering "unnecessarily" in its pursuit' (1999: 201). With the emergence of radical asymmetry, we are confronted by the inverse of this maxim: that our end is virtuous *on account* of our lack of suffering.[11]

It is first worth stating the obvious; military and moral superiority are two very different things. It is partly for this reason that Shue writes:

> War is not a distributive mechanism. War is essentially not a mechanism that can allot the evils of death and destruction only, or even mainly, to those who have done wrong and protect the lives and goods that belong to those who have done right. (2008: 102)

Within conditions of radical asymmetry, however, war *does* transform into a 'distributive mechanism', at least for the stronger party. This highlights one of the broader implications of radical asymmetry for the regulation of lethal force. Its distortion of the distinction between war and more punitive, judicial violence has the potential to lead to the moral devaluation of those targeted. This, in turn, makes an erosion of essential limits on violence more likely. Coates identifies this as one of the dangers associated with the subordination of the Just War criterion of *in bello* to *ad bellum*. 'Unilateral justice', he argues, risks:

> [U]ndermining those moral limits that the [Just War] tradition seeks to place on war. It is just such an assumption that lies behind the holy war in its religious and secular or ideological forms and that leads in extreme, but by no means rare, cases to a war of annihilation. (Coates 1997: 151)

These risks can be witnessed within the Just War Tradition as early as Augustine, who, in viewing war through a lens of divine justice, gave sanction to a prosecution of Christian violence that was often limitless (Clark 1988: 37). Today, 'unilateral justice', along with its attendant problems, finds its clearest expression in the radically asymmetric, UAV-exclusive violence of the United States. Here the work of Bauman is instructive, particularly his

[11] Like most changes in war, this is not entirely without historical precedent. In an article titled 'Exterminating Angels', authors Aksoy and Robins illustrated the extent to which technological proficiency during the First Gulf War reinforced American claims of moral legitimacy, as well as a broader narrative of Western cultural and moral superiority (1991). For earlier examples of this same logic, see Adas (2014) and Ellis (1986).

analysis of the intimate link between modernity and genocide. In *Modernity and the Holocaust*, Bauman writes the following:

> *Modern genocide is genocide with a purpose.* Getting rid of the adversary is not an end in itself. It is a means to an end: a necessity that stems from the ultimate objective...*of a better, and radically different, society*...This is a gardener's vision, projected upon a world-size screen...Some gardeners hate the weeds that spoil their design – that ugliness in the midst of beauty, litter in the midst of serene order. Some others are quite unemotional about them: just a problem to be solved, an extra job to be done. Not that it makes a difference to the weeds; both gardeners exterminate them. If asked or given a chance to pause and ponder, both would agree; weeds must die not so much because of what they are, as because of what the beautiful, orderly garden ought to be. (1989: 91–2)

Framed in these terms, violence becomes not a destructive act, but rather a creative process of disinfection. Terrorists are mere hindrances to be got rid of; cleared away. They are dehumanized. The obvious should first be acknowledged here. Radical asymmetry, as it is currently imposed, is in no way genocidal, neither in intent nor practice. The claim here is a more modest one. This chapter has already argued that radical asymmetry calls into question the permissiveness of violence by the stronger party. This same imbalance has the potential to jeopardize a more basic connection between enemies in war, grounded in the mutual recognition of the other's moral value.

Despite humanity's long-term transition towards peace and for all our noble qualities, we as a species continue to bear what Charles Darwin referred to as 'the indelible stamp of our lowly origin' (1889: 619). Of these less desirable traits, perhaps the most harmful is our enduring capacity for cruel and destructive conduct. Central to the pacification of humans has been the steady taming (through morality, law, community, culture, etc.) of these baser qualities. Yet it should not be forgotten that civilization has restrained these impulses only, not eradicated them. This is why war is a terrible thing. In war, these chains, which have taken the full span of human history to craft, are struck. In his influential work, *The Civilizing Process*, Elias wrote of the taming of impulses, such as 'cruelty and joy in the destruction and torment of others', through 'increasingly strong social control anchored in the state organization' (1994: 192). The fragility of this control is exposed 'only at times of social upheaval [war] or where social control is looser [the frontier]' (Elias 1994: 192).

War upends the rules and standards of normal society, replacing them with an escalatory dynamic of bloodshed—man-*in extremis.* According to Clausewitz, reciprocal escalation meant that war was a phenomenon in which no logical limits could be placed on the act of force (2008: 15). Carl Schmitt echoed this, writing of the 'vicious circle of reprisals and anti-reprisals' that characterized warfare (2007: 28).

The warrior ethos and Just War Tradition are both attempts to forestall this escalation, by 're-chaining' fighters to a fixed set of rules. Yet, even at their most effective they do so imperfectly. The all too frequent descent of war into atrocity and massacre throughout history is testament to the ruinous consequences of this spiralling effect.

In light of these dangers, there are few things to be more valued in war than the preservation of bonds of humanity between those that seek to kill one another. There is great worth in the basic recognition that those we harm are fellow moral agents—or put more simply, 'poor devils like us' (Remarque 1929: 223). Observers of war have long maintained the importance of this standard. Vattel counselled military leaders to 'never forget that our enemies are men' (2008: 563). Nagel agreed, citing the importance of 'interpersonal' relationships in wartime (1972: 136).

When this bond of shared humanity and victimhood between enemies is sundered, an erosion of behavioural restraint often follows. In his book, *Less than Human*, Smith argues that 'dehumanization' is the fundamental component in all atrocity. He goes on to demonstrate this in relation to the historical status of Jews, sub-Saharan Africans, and Native Americans (2011). Further examples of this in the context of Western warfare can be located throughout history, up to and including the twentieth century. Whereas the violence between 'civilized' nations underwent a gradual process of 'humanization', warfare waged beyond these conceptual boundaries has, until relatively recently, been governed by few constraints (Best 1980: 217; Glover 2001: 28; Howard 1994: 5; Lindqvist 2001: Sec. 30).[12]

It is certainly correct that the mutual violence of the traditional battlefield can exacerbate feelings of hatred between enemies and lead to less restricted warfare. The argument here is that a comparable or potentially even greater

[12] The consequences of denied humanity on the battlefield were highlighted in stark terms during a range of historical episodes, including the Greek/Persian wars (Ober 1994: 18); European colonial policing of Africa (Hardinge, cited in Ellis 1986: 92; Fuller 1923: 191–2; Reeves 1923, cited in Colby 1927: 279); US/Japanese hostilities during the Second World War (Dower 1987: 77–96); Nazi conduct on the Eastern Front during the same conflict (Hitler, cited in Lindqvist 2001: Sec. 180); and American excesses during the Vietnam War (Paskins and Dockrill 1979: 204).

erosion of restraint can result from the military imbalance of radical asymmetry. The preservation of essential limits on war 'is inextricably linked to the enemy's inclusion within... the sphere of equal moral standing' (Erskine 2008: 195). Radical asymmetry has altered the nature of war to a degree that imperils this inclusion, pushing an increasingly dehumanized enemy further outside the moral 'zone of application' (Honneth 1995: 113). Carl Schmitt outlined the dangers of this exclusion in the concluding chapter of *The Nomos of the Earth*:

> If the weapons are conspicuously unequal, then the mutual concept of war conceived in terms of an equal plane is lacking. To war on both sides belongs a certain chance, a minimum of possibility for victory. Once that ceases to be the case, the opponent becomes nothing more than an *object of violent measures*. (2006: 320, my emphasis)

In situations such as these, with the stronger party no longer bound by a 'dialectic of killing and dying' what remains is a mode of violence that more closely approximates 'certain kinds of pest control' (Münkler 2003: 132); a gardener's vision of war.

To be clear, radical asymmetry is not the sole explanation for this erosion of enemy status. There are a range of factors that explain the increasing tendency of the United States and other Western powers to criminalize and dehumanize those they war against. From the 1990s onwards there has been a concerted effort among Western states to outlaw and pathologize war, while at the same time legitimizing its use in the name of humanity (Shaw 2005: 92; Werner 2004). This general trend intensified following the attacks of September 11. The War on Terror has been the catalyst for the realization of many of Schmitt's fears, most notably, that war as a symmetrical confrontation between 'just and equal enemies', would be replaced by punitive judicial measures against enemies reconceived as 'delinquents, troublemakers, pirates, and gangsters' (C. Schmitt 1999: 60–1).[13] What is argued here is that the profound disparity in capability, threat, and risk between the United States and those it targets has amplified and accelerated this trend.

[13] '[T]here is the danger', writes Linklater, 'that [the liberal peace] will be grounded in a sense of superiority to the non-liberal world, in a lack of interest in the suffering of non-liberal peoples and in the belief that the principles that apply in the liberal world do not apply in wars with illiberal states' (2002: 333). See also, Doyle (1983); Goldgeier and McFaul (1992); Howard (1989); and Sadowski (1998).

In the context of UAV-exclusive violence, radical asymmetry manifests as an increasingly sterile, bureaucratized, and detached mode of American killing (Becker and Shane 2012: A1; Obama, cited in Chait 2016). This detachment has translated into a dehumanization by the United States of those it targets (Chamayou 2015: 165–6; Clark 2015: 116; Gholiagha 2015; Hass, cited in Pilkington 2015; Muñoz-Rojas and Frésard 2004: 199; Welsh 2015: 43; Wilcox 2015: 151; Zenko 2013b). In some cases, this has led to a more severe decline in the targeting practices of the United States, from morally dubious acts to those that skirt the very edge of outright moral and legal violation (Becker and Shane 2012: A1; Casey-Maslen 2014: 397; Gholiagha 2015: 140; Kaag and Kreps 2014: 32–3; Scahill 2015; Stanford Law School 2012: 33). This suggests that in addition to challenging the coherence of the moral frameworks of war, radical asymmetry incentivizes the adoption of new and problematic battlefield practices. This breakdown of restraint between warring parties is further compounded by the weaker party's limited scope to retaliate.

Radical Asymmetry and the Redirection of Risk

When outlining his theory of special relativity, Einstein observed that energy cannot be created or destroyed; it can only be changed from one form to another. When conceptualizing reciprocal risk in war, it is useful to imagine something analogous to the law of conservation. As noted earlier in this chapter, weaker parties to a conflict have long sought to offset their military inferiority through the utilization of unconventional tactics. This may include conduct that pressures or even outright violates the laws of war. Technologically disadvantaged parties have at times utilized human shields or misused specially protected objects—launching attacks from hospitals, schools, or buildings of cultural and spiritual significance (M. Schmitt 2006: 152–3). Some conventionally weaker parties have gone further, redirecting their violence away from the military targets of the strong and towards its civilian base.

Radically uneven ratios of military power increase the likelihood of this (Kahn 2002: 6; Osiel 2009: 202–3). When assessing the implications of radical asymmetry for the regulation of lethal force, the significance of this redirection must be acknowledged. The physical dislocation of humans from the battlefield by the United States has narrowed the scope of legitimate retaliation among those targeted to an unprecedented degree. Within such conditions, the weaker side may then frame war crimes as the only possible recourse short of surrender or total defeat (Killmister 2008: 122; Turns 2014: 213).

This should not be read as an endorsement for a relaxation of *in bello* restraints for weaker parties, even for those subjected to the most profound of risk disparities in war. As a number of authors have outlined, no degree of 'unfairness' in battle excuses the deliberate targeting of civilians (Roberts 2008: 237–8; Shue 2013: 287–92; Statman 2012: 108–9; Walzer 2013b). Consequentialist objections aside, there exist a wide range of deontological reasons to reject justifications for terrorism, the most convincing of which stem from Kant's categorical imperative. There is a transcendent value to people, and as such, they are not to be treated merely as means, but also ends, and subjected to harm only when liable to such a fate. The blanket immunity of non-combatants is unaffected by the military advantage of one side. The rights of civilians are simply universal human rights. Terrorism, when directed against the innocent and non-threatening, is, by definition, *prima facie* wrong.

It is incumbent upon those committed to preserving these protections, however, to seriously engage with the factors that incentivize deviation. The radical asymmetry imposed by the United States has immunized its military targets from retaliation to a historically unprecedented degree. In the process, it has made more likely the displacement of a commensurate degree of risk onto those upon whose behalf they fight.

As outlined in the previous chapter, one of the main criticisms directed against radical asymmetry is that it is too often obtained at the expense of the safety of enemy civilians. This reflects a long-standing debate within the Just War Tradition and laws of war regarding the level of risk combatants must themselves accept in order to minimize harm to the non-combatants of the opposing side. This debate is roughly divided into two groups. The first insist that combatants must incur additional costs, in the form of increased personal risk, in order to enhance the safety of enemy non-combatants. Walzer, among others, sought to address this with an amendment to the Doctrine of Double Effect (DDE). According to Walzer, for DDE to be permissible, military attacks must have a 'double intention'—the first, that enemy civilians not be the intended target, and the second, 'that the foreseeable evil be reduced as far as possible' (2006a: 155).

The second group argues against this, suggesting that such measures undervalue the safety of combatants (Cohen 1989: 3; Kasher and Yadlin 2005: 14–15; Porat and Ziv 2014). Despite these differences, both positions align regarding the expectation that militaries assume risk to protect their *own* non-combatants. States maintain a strong duty of care to their soldiers, and must mitigate expected incoming harms wherever possible. This obligation, however, remains secondary to the fundamental duty of the state, which is to secure its own citizens from violent harm. A state that neglects this moral duty undermines its own *raison d'être*.

In our current context, one might rightfully respond to these arguments by highlighting the moral crimes of the range of actors currently targeted by the United States through radically asymmetric means. This is a reasonable claim; ISIS and others need no additional incentive to redirect their violence against Western non-combatants. One should not, however, overlook the near-certainty that in the coming years, more and more states will utilize the robotic weaponry that enables such asymmetry. The moral justifications offered for these future acts of violence are likely to draw upon the precedents that are currently being established by the United States. Problematically, many of these precedents seem likely to enhance, rather than mitigate, excessive and unjustified violence.

Conclusion

This chapter has outlined the main contentions of this book. It has argued that the permissiveness of the moral standards of inter-combatant violence derives its coherence from an assumption of war as a site of reciprocal structural risk. The potential of radically asymmetric violence to undermine this assumption, and with it, the moral premise of the permissiveness itself, represents a profound challenge to the existing justifications for killing in war. This challenge goes beyond the warrior ethos, calling into question the very ability of the *in bello* frameworks of war to properly designate liability to harm.

Radical asymmetry has further implications in regard to the regulation of lethal force. First, the one-directional nature of its violence makes a dehumanization of the weaker party more likely. In addition, radical asymmetry incentivizes those targeted to redirect violence from the military to the civilians of the stronger side. Both outcomes place further strain on battlefield restraint.

Having summarized the overall argument of this book, the next five chapters will explore the dilemma of radical asymmetry in greater theoretical and empirical depth. This will begin with an examination of the role of reciprocity within the warrior ethos, and specifically, its justification for combatant violence.

PART II
RECIPROCAL RISK IN WAR

3
Reciprocity and the Warrior Ethos

Introduction

Chapter 2 argued that radically asymmetric violence represents a profound challenge to the moral justifications for killing in war. This challenge is qualitatively distinct, both in significance and implications, from the ethical tensions generated by past military imbalance. This distinction is grounded in the potential of radical asymmetry to undermine the *structural* reciprocity of war, and with it, the very coherence of the permissive moral standards of inter-combatant violence. These findings will be confirmed in the following two chapters, which locate the role of reciprocal risk in the warrior ethos and Just War Tradition. Both chapters will substantiate the crucial difference in war between the *ethical* challenge of non-risk and the *moral* challenge of non-threat, as well as the moral distinction between inter-combatant and inter-belligerent asymmetry.

Chapter 3 begins this analysis with the warrior ethos. This code of martial honour has historically operated as a kind of socialized motivation, created and maintained to safeguard the ethical status of the warrior. Within this framework, ethically justified violence has long been preconditioned on a reciprocal imposition of inter-combatant risk. Adherents to the warrior ethos have, in turn, historically disdained killing in the absence of meaningful physical vulnerability. As this chapter will further argue, however, another key aspect of the warrior ethos is its capability to adapt in response to changes in the material and ideational character of war. An important consequence of this adaption has been the gradual de-emphasis—though not elimination—of the role of physical risk in the ethical justification for killing in war. Though personal, physical risk endures today as a determinant of battlefield legitimacy, it is increasingly viewed as a secondary consideration to 'combat responsibility'—typically gauged in reference to a fighter's professionalism and adherence to the explicit rules and standards of war. This is significant, as it illustrates that the challenge of radical asymmetry, at least in terms of its tension with the warrior ethos, is a resolvable one.

Asymmetric Killing: Risk Avoidance, Just War, and the Warrior Ethos. Neil C. Renic, Oxford University Press (2020).

DOI: 10.1093/oso/9780198851462.001.0001

This chapter first examines the 'Age of the Warrior', focusing on the military ethos of the ancient Greeks and Romans, as well as the medieval code of chivalry. Across these periods, reciprocal risk between individual adversaries was a dominant factor in determinations of legitimate violence. The chapter will then shift forward in time to address the 'Age of the Soldier', examining both early modern and modern war, as well as the emergence of 'post-heroic' warfare. It argues that although reciprocal risk has endured across these periods as an indicator of ethical killing, it is increasingly counterbalanced, and superseded, by an expectation of restrained, humane, and professional conduct among fighters.

The Age of the Warrior

This section explores the fundamental role of reciprocal risk within early iterations of the Western warrior ethos. Over the course of this analysis, two important distinctions will be highlighted between the ethical and moral conception of reciprocal risk. First, in contrast to the Just War Tradition, the warrior ethos has traditionally grounded its justification for lethal force in an assumption of inter-combatant (individual), rather than inter-belligerent (collective) reciprocal risk. Second, and in further contrast to the moral frameworks of war, the warrior ethos has traditionally regarded the *risk* assumed by the fighter, rather than the *threat* projected by the target, as the critical factor in permissible killing. This will be demonstrated through analysis of ancient Greek and Roman war, as well as the medieval code of chivalry. Central to each of these versions of the warrior ethos was the recognition that battlefield honour could be accrued by fighters only when their violence was accompanied by some degree of physical vulnerability.

Ancient Greek and Roman Warfare

Greek Warfare

Any inquiry into the historical role of reciprocal risk in the warrior ethos must begin with the ancient Greeks, for it is within this period that one of the earliest expressions of the Western way of war can be found. Hanson traces this link back to the Hoplite warrior—citizen-soldiers of ancient Greek city-states—who in war famously adopted the Phalanx, a rectangular mass infantry formation (2001). For others, the focal point of this Hellenic legacy is located in the fictional works of Homer. Wheeler highlights the historical influence of the 'Achilles ethos...the advocacy of chivalry, face-to-face confrontation, open battle, and use of force' (1990: 124).[1] Within both these representations of war and the warrior, reciprocal risk played an important function in the ethical justification for lethal force.

It would be incorrect to characterize the Greek relationship with the battlefield as one of pure glorification. Greek philosophy and literature examines at length the cruel, tragic, and wasteful qualities of war. Yet at the same time, the Greeks regarded war as an existential and transformative experience; a 'master text' by which the warrior could better know himself (Coker 2002b: 58).[2] War, like no other setting, exposed the defining quality of man, be it cowardice or courage; avarice or self-sacrifice. The unveiling of this quality, however, was contingent on a warrior's engagement with the fear and danger inherent to violent combat. Ultimately, it was the acceptance of physical risk in battle that transformed the warrior experience from an instrumental practice to a 'humanistic endeavour' (Coker 2002b: 17).

This is evident within ancient Greek depictions of the ideal warrior archetype. Describing what makes a good Hoplite, the seventh-century Lyric poet Tyrtaeus wrote, '...no man ever proves himself a good man in war unless he endures to face the blood and slaughter' (cited in Lattimore 1955: 14). Admiration is reserved here only for those warriors that expose themselves to the physical risk of battle. Contemporary scholars such as Keegan reaffirm this, highlighting the ability of the Greeks to transcend the biological resistance to 'pushing personal exposure to its intolerable limits' (1994: 251). 'They bore it', suggests Keegan, 'only because all shared the risk equally...' (1994: 251).

'To the Greeks', Hanson argues, 'anything less than a "fair" fight – that is, a daylight clash of two massed phalanxes – was no fight at all, however decisive' (1989: 14). Depicted here is a mode of warfare governed as much by ritual and symbolism, as rationality, with primacy given to the *act* rather than utility of war. Hanson goes on to recount various examples in which Hoplite armies emerged from fortifications or sacrificed high ground, not for any tactical advantage, but because they had internalized a code of military behaviour that privileged 'the risk of death in a short battle' over protracted and

[1] See also, Gusterson (2014: 191–3).

[2] The existential nature of war has been a recurring theme throughout history. Morison described military service as a realm wholly separate from normal society, within which combatants were able to 'find the definition of their whole being' (1966: 34). Broyles echoed this, describing war as 'the only way in which most men touch the mythic domain of their soul' (cited in Mueller 2004: 10). For similar accounts, see Gray (1998); Nadelson (2005); and van Creveld (1991).

indecisive engagements (1992: 188). Herodotus' account of the Battle of Marathon offers a stark illustration of this preference. According to Herodotus, the Persian commander Mardonius said to his king:

> ...these Greeks are wont to wage war against one another in the most foolish way through sheer perversity and foolishness. For no sooner is war proclaimed than they search out the smoothest and fairest plain that is to be found in all the land and there they assemble and fight... (cited in Dawson 1996: 47).

The Greek regard for reciprocal risk was reflected not only in the battlefield practices they adopted, but also in those they denigrated. In particular, the Greeks disdained modes of warfare that elevated deception and subterfuge at the expense of direct exposure to physical risk. According to Euripides: '[a] brave man thinks it unworthy to kill his enemy by stealth; he meets him face to face...do not praise the clever spear of one who *steals* victory' (cited in Whetham 2005: 2, emphasis in original). Plato resented naval warfare for similar reasons, believing it incentivized warriors to hit and run, rather than stand and fight (1988: Book IV). As both accounts clarify, the Greeks held victory in lower regard if obtained in the absence of personal risk.

Further contempt was reserved for those who fought and killed from afar. This stigma was attached to the slinger, the javelin thrower, and most of all, the archer: 'all men who could kill "good" infantry...with little risk to themselves' (Hanson 1989: 15). For Euripides, the bow was 'no proof of manly courage; no, your real man stands firm in the ranks and dares to face the gash the spear makes' (cited in Robinson 2006: 28). This again reinforces the centrality of reciprocal, inter-combatant risk in the Greek warrior ethos:

> The Greeks had a determined disdain for remote forms of warfare...Greek warriors found the measure of themselves and of others in close combat...Killing remotely, without risk, was not what war was about; it was instead, *phonos* – slaughter or murder – and there was no honour to be found nor glory to be won in that. (Meagher 2014: 28, emphasis in original)[3]

Reciprocal risk was more than simply a preferred feature of Greek warfare. Rather, it stood as a principal determinant of the ethical legitimacy of

[3] See also, Cartledge (2002: 35); Hartog (1988: 45); Homer (2003:191); Garlan (1975: 128); O'Driscoll (2015: 5–6); Pritchett (1974; 174); and Rawlings (2007: 82–3).

combatants.[4] The importance of reciprocity could also be witnessed in the Greek treatment of the dead following battle. Socrates considered the plundering of the enemy dead 'demeaning', 'mercenary', and 'womanish', arguing that it gave 'cowards an excuse not to go after those who are offering resistance' (Plato 2000: 170).

This is not to argue that all Greek warfare adhered to a strict equality of risk. Alongside the Hoplite Phalanx, Greek armies eventually took advantage of archers, cavalry, light-armed javelin throwers, and marines. Furthermore, there are numerous accounts in the historical record of Greek raids and ambushes. Wheeler terms this rival aspect of Greek warfare, centred more on deception than courage, the 'Odysseus ethos' (1990: 124). Crucially though, these exceptions do not undermine the role of reciprocal risk, in either the Greek or subsequent conceptions of the warrior ethos. Sidebottom suggests that rather than view the Western way of war as a genuine continuity of practices, it is better to regard it as an ideology, or ideal, founded by the Greeks (2004: preface). Within this ideology, reciprocal inter-combatant risk was fundamental to the delineation of ethical conduct, regardless, and often in spite of actual practice.[5]

As Western warfare evolved, so too did the attitudes and conduct of its participants. What endured, however, was an assumption of reciprocal risk within the warrior ethos. This will be illustrated through an examination of ancient Roman war.

Roman Warfare

The warfare of the ancient Romans was marked to a significant degree by the growing ascendancy of instrumentalism over humanism in battle.[6] Keegan

[4] There is significant debate regarding the extent to which opposition to 'riskless' killing translated into formal and enforceable prohibitions. Polybius does cite one convention outlawing long-range missile weapons (2003: 415). This was likely in reference to a treaty between the warring Greek powers of archaic Chalcis and Eretria; an agreement that some contemporary scholars have called into question (Wheeler 1987: 157). Greek warfare was guided for much of its history by a body of unwritten conventions, commonly referred to as the 'common customs of the Hellenes'. This framework prohibited the direct targeting of non-combatants and prisoners of war (Christopher 2004: 10; Ober 1994: 13; Van Damme 2002: 155). Given their informal nature, conformity with these standards derived primarily from a fear of reputational damage. For contrasting views on the Greek treatment of prisoners of war, see Ducrey (1968, cited in Patterson 1982: 110); and Lanni (2008: 474).

[5] Within the Greek context, this ideology was strongest between 700 and 450 BCE. The social and political displacement of the Athenian Hoplites, resulting from the combined growth of democratic political culture and the navy, were a major factor in the erosion of the regulatory frameworks of war after this period (Ober 1994: 20). In particular, the Peloponnesian Wars triggered a number of social and strategic changes to warfare, which together 'reduced the centrality of the warrior class upon which many traditional martial customs rested' (Bellamy 2006: 16).

[6] Some scholars appeared to lament this fact. Polybius, for example, said in reference to the Greeks: 'the ancients chose not to conquer their enemies by deception...therefore they declared wars in

argues that no military before Rome (nor any for a significant period after) achieved its level of bureaucratically regulated recruitment, supply, organization, and command (1994: 267). Roman warfare was equally unmatched in terms of its violence, which even when viewed alongside the conflict of its age, stood apart in its lack of restraint (Harris 1979: 56). Medieval theorists regarded Roman violence as so excessive, they developed the term *Bellum Romanum*, a category of warfare in which the distinction between combatants and non-combatants was erased and indiscriminate slaughter permitted (Stacey 1994: 28). This was contrasted with *Bellum Hostile*—conflict conducted between moral equals. Yet even within Roman warfare, a commitment to reciprocal risk can be found within the warrior ethos.

This can first be observed in the works of Marcus Tullius Cicero (106–43 BCE). Cicero's adage, *Inter arma enim silent leges*—'in times of war, the laws fall silent' (Reichberg, Syse, and Begby 2006: 50)—has long been cited as a validation of the realist approach to wartime conduct. A closer reading of Cicero, however, reveals a deeper commitment to battlefield restraint, one that was underpinned by a shared code of martial honour. Cicero advanced a Stoic philosophical outlook that regarded lapse of character and virtue as the greatest loss that could be suffered (French 2003: 66). Combatants were to behave honourably on the battlefield, even towards those they sought to kill. Cicero disdained the use of poison in war for this reason. Recounting a conflict between Rome (led by Caius Fabricus) and the forces of King Pyrrhus of Epirus, Cicero wrote:

> For when King Pyrrhus had made aggressive war upon the Roman people…a deserter from him came into the camp of Fabricus and promised…[to] return secretly into the camp of Pyrrhus, and dispatch him with poison. Fabricus took care that this man should be sent back in custody to Pyrrhus, and this conduct of his was applauded by the senate…it would have been a great disgrace and scandal, that he with whom the contest was for glory, had been conquered, not by valor, but by villainy. (1850: 151)

This echoes a sentiment found throughout the history of war, regarding the uniquely dishonourable nature of poison. Italian jurist Alberico Gentili (1552–1608) referred to Roman prohibitions when condemning the use of weapons that 'shame the steel with venom' as 'abominable' and 'contrary to the laws of the Gods' (1933: 156). Gentili went on to recommend that the

advance, announcing when and where they were going to deploy. But now they say only a poor general does anything openly in war' (cited in Krentz 2000: 168).

prohibition against poison endure, 'because war ought to be limited to things which it is within human power to resist' (1933: 158). Hugo Grotius also wrote on the use of poison in war. Though he conceded its permissibility in strictly natural law terms, he nevertheless viewed it as underhanded. The reason for this was that it targeted the enemy without his knowing, and thus deprived him of the 'power of defending himself' (2012: 353).

It is important to note that while the use of poison has long been condemned within both the warrior ethos and Just War Tradition, the grounds for such a prohibition has differed between the frameworks. Roman opposition to the use of poison can be understood as a reaffirmation of martial virtue. In contrast to the face-to-face violence of the battlefield, in which a combatant risked his own life to defeat the enemy, killing through the use of poison required no courage and thus could bestow neither honour nor glory.[7] Hallissy highlights the importance of reciprocal risk in this ethos-based objection to poison:

> Poison can never be used as an honourable weapon in a fair duel between worthy opponents, as the sword or gun, male weapons, can . . . the dueller is open, honest, and strong; the poisoner, fraudulent, scheming, and weak. A man with a gun or a sword is a threat, but he declares himself to be so, and his intended victim can arm himself: may the best man win . . . (1987: 5–6)

A key aspect of the warrior ethos is its inward-focus; combatants have a duty to themselves to maintain their honour and physical courage no matter the challenge set before them. The Just War Tradition, in contrast, though equally critical of poison, shifts its emphasis to the moral status of the target. This shift can be observed in the work of Anscombe, who wrote:

> [I]f I push a man over a cliff when he is menacing my life, his death is considered as intended by me, but the intention to be justifiable for the sake of self-defence. Yet the lawyers would hardly find the laying of poison tolerable as an act of self-defence, but only killing by a violent action in a moment of violence. (2006: 628–9)

This reaffirms a broader contention of this chapter. Killing in the absence of personal, physical risk has throughout history been argued to misalign with the warrior ethos. *Non-risk* is the nucleus of this ethical tension. The

[7] The Roman veneration of martial virtues also translated into a broader stigmatization of assassination (Thomas 2001: 52–3).

fundamental consideration of Just War theory, in contrast, is the status of the potential targets of such violence. The right to kill within these moral frameworks is brought into question only upon a determination of *non-threat*.[8]

To return to the Roman context, this ethical/moral divide can again be witnessed in relation to the treatment of defenceless captives in war. *Rhetorica ad Herennium,* a Roman work formally attributed to Cicero, but in fact of unknown origin, argued against the putting to death of those 'captured by force of arms'. The reason for this was that 'it is the characteristic of a brave man to regard rivals for victory as enemies, but when they have been vanquished to consider them as fellow men' (1954: Book IV). This appeal makes no reference to the inherent rights of the captive, but rather the ethical importance of courage. Reinforcing this distinction between agent non-risk and target non-threat is critical when determining the significance and implications of emerging military imbalance. Modes of asymmetric violence that imperil the former, but leave the latter largely intact, are indeed a challenge, but crucially, a challenge only to the warrior ethos.

The ancient Greeks and Romans notably diverged in their approach to war. The militaries of both, however, had internalized a warrior ethos that considered violence worthy of esteem only when exercised within conditions of reciprocal risk. This chapter will now turn to arguably the most celebrated period in the history of the Western warrior ethos, the age of chivalry. During this time, reciprocal risk remained a critical determinant of the ethical status of fighters.

Medieval European Warfare

The frameworks that regulate contemporary warfare are, to a significant extent, outgrowths of the rules and standards that developed during the medieval period of Europe. The two major sources of these rules were the Christian Just War Tradition and European code of chivalry (Solis 2010: 5). The following section will outline the role of reciprocal risk within the latter framework. This analysis further clarifies the distinction between the individualized and *risk*-based conception of reciprocity that underpinned the warrior ethos and the *threat*-based reciprocity of the Just War Tradition.

[8] As Chapter 4 will clarify, the precise role of material threat in the moral justification for inter-combatant violence differs between the three most dominant accounts of Just War.

The warrior ethos arguably found its fullest expression in Western warfare during the European Age of Chivalry, generally recognized to have taken place between the twelfth and sixteenth centuries (Keen 1984: 238). This code of conduct was grounded in the warriors' own self-realization and functioned to channel, constrain, and legitimize the violent interactions between opponents in battle. Just as with Greek philosophy and Roman Stoicism, the European code of chivalry hinged on the conviction that there were experiences in battle worse than death, such as betrayal of martial honour. Though chivalric codes themselves varied between European regions, all esteemed a similar set of virtues, one of the most fundamental of which was physical courage: 'for chivalry abideth not so agreeably in no place as in noblesse of courage' (Keen 1984: 10).

Determining the extent to which this code translated into explicit restraints in battle is difficult, as few periods have been so heavily mythologized. It does appear, though, that chivalry functioned to at least partly constrain violence between combatants, while redirecting it away from the harmless. Kennedy writes that throughout this period, terms such as 'treason' and 'murder' were used to describe violations of chivalry, including the use of violence against surprised or otherwise disadvantaged individuals (cited in Ackerman 2003: 126). Implicit in this condemnation was a conviction that for killing to be honourable the warrior must accept a minimum degree of physical risk. Demonstrating this further was the long-standing requirement that knights avoid harming women, children, 'and all the weak and helpless' (Painter 1969: 79).

It is again worth reiterating the distinction between the warrior ethos and Just War Tradition regarding the role of reciprocal risk. Just War theory, expressed during this period through Canon law, regarded immunity from direct and deliberate harm as a *right* held by non-combatants. This is evidenced in the spiritual sanctions sought during the tenth-century 'Peace of God', and mid-eleventh-century 'Truce of God' movements by the Catholic Church. The main tenets of both were summarized at the Second Lateran Council in 1139. In reference to the principle of discrimination, the Council decreed that 'priests, clerics, monks, pilgrims, merchants, and peasants . . . [must] be left in peace at all times' (1139: Canon 11). It was permissible to kill only those enemies that bore arms, and as such, represented a material threat (Howard 1994: 8; Parker 1994: 41).

By the end of the medieval period, this list had expanded to include children, women, the infirm, the elderly, and those of unsound mind (Johnson 1981: 132). Francisco de Vitoria (1492–1546) argued in subsequent years for similar restraints. According to Vitoria: 'we may not use the sword against those who have not harmed us; to kill the innocent is prohibited by natural law' (1991: 304). Vitoria extended these protections to women, children and 'harmless agricultural folk', as well as enemy soldiers no longer capable of causing damage (cited in Hartigan 1982: 87). Francisco Suárez (1548–1617) offered a similar class of non-combatants, including 'all unable to bear arms' (2006: 363).

For the church, along with subsequent Just War thinkers, non-combatant immunity was an inherent right, grounded in an individual's inability to harm (Dill 2015b: 115; Kenny 1985: 10). As this re-establishes, within the context of the *moral* frameworks of war, the elimination of physical risk by one side matters only to the degree that it invalidates the *threat* of those targeted.

The code of chivalry also prohibited the lethal targeting of the defenceless. In an important contrast, however, its practitioners regarded these protections not as a right held by the vulnerable, but rather a *gift* granted to the weak by the strong and superior (Johnson 1981: 138–40). This distinction would endure until the fourteenth century when church and chivalric justifications for non-combatant immunity merged (Johnson 1981: 138–40). In reference to latter, Coker observes that '[t]raditionally, the death of civilians has offended moral sensibilities, not because they were considered more innocent or helpless but because no courage was involved on the warrior's part' (2001a: 37).

Conformity with non-combatant immunity served as a reaffirmation of the warrior's own martial standing. This, along with other restrictions, provided 'a kind of... [ethical]... and psychological armour that protect[ed] the combatant from becoming a monster in his or her own eyes' (French 2003: 7, 10). The price paid for this ethical imprimatur was some measure of physical vulnerability. Without risk, the warrior could not earn the *privilege* to lethally target their opponent. This was further detailed in the works of French writer Christine de Pizan (1364–1431 CE). Within *The Book of Deeds of Arms and of Chivalry*, Pizan wrote: 'what honour can accrue to a prince in killing, overrunning, or seizing people who have never borne arms nor could make use of them...' (1999: 171).[9]

The importance of reciprocity was also evident throughout this period in the opposition that emerged to modes of violence that mitigated physical risk to the operator. Arguably the most famous medieval example of formal opposition to 'unfair' weapons can be found in the Second Lateran Council of 1139. Pope Innocent II prohibited 'under anathema that murderous art of

[9] Pizan did accept that non-combatants could be injured and killed indirectly during the course of war, but stressed the need for combatants to take every precaution to prevent their death.

crossbowmen and archers, which is hateful to God, to be employed against Christians and Catholics...' (Second Lateran Council 1139: Canon 29).[10] This was reiterated by the Third Lateran Council, as well as the 1500 *Corpus Juris Canonici*, which forbade the use of darts, arrows, or catapults (Green 1998: 46). Though quickly disregarded, these prohibitions still highlight the degree to which missile weapons were perceived to deviate from the ethical ideal of warrior violence.

The extent of this deviation was brought into stark relief on the battlefield itself. French knight Bayard justified the execution of captured crossbowmen on the grounds that 'their weapon was a cowardly one and their behaviour treacherous' (Keegan 1994: 333). Such was the contempt directed towards ranged troops that when French knights charged against the English line at Crécy, they 'literally rode down their own crossbowmen' (Stacey 1994: 36).[11]

The code of chivalry, along with earlier iterations of the warrior ethos, occupied a distinctly different regulatory space than the Just War Tradition. The principles of Just War were developed and imposed from outside the battlefield, with an emphasis on the potential victims of war. The warrior ethos, in contrast, was a creed and culture created and fostered by the agents of violence themselves, within the crucible of battle. Adherence to this ethical framework provided the means for combatants to elevate their violence, and by extension, themselves, to something more noble. In this pursuit, a reciprocal imposition of risk between warrior and target has long been regarded as an essential condition in the ethical justification for killing.

Judged in these terms, it seems reasonable to conclude that any emerging mode of violence that sufficiently minimizes physical risk to the warrior will struggle to align with the ethical virtues of the warrior ethos. Courage in the face of the enemy presupposes an active contestation of force between opponents, one of the possible outcomes of which may be a violent death. And as Anscombe rightly questioned, 'how can you be courageous... without having reason to think you are in danger[?]' (1981a: 64). However, in order to determine the extent to which military imbalance, including contemporary conditions of radical asymmetry, represents a challenge to the ethical justification for killing in war, it is necessary to also examine the inherent adaptability of

[10] In the case of missile weapons and early firearms, resistance was rarely based exclusively on the perception of an unjust asymmetry. Opposition to both longbows and crossbows was also grounded in the belief that the wounds they delivered were particularly 'barbarous', as well as the conviction that such weapons empowered the common fighter and thus threatened the privileged status of the warrior class (Boothby 2009: 373; Hatto 1940; Payne-Galloway 1903: 3).

[11] Differences in social status, as well as weaponry, likely played a role here.

the warrior ethos. As this chapter will next illustrate, the warrior ethos, rather than remaining static, has evolved in response to the modernization of war. One consequence of this evolution has been a gradual de-emphasis (though not elimination) of the role of reciprocal physical risk in the ethical status of fighters.

The Age of the Soldier

When one imagines an army, it is likely the material components that first come to mind; the uniforms, the weaponry, and most importantly, those crucial individuals whom we task with the killing and subject to the dying. Yet an army is also a social construction, grounded in shared values and expectations. These include formal regulations, written orders, and a defined hierarchical authority, as well as military traditions, collective narratives, and a binding ethos (Shay 2003: 6). As this chapter has so far outlined, throughout most of the history of Western warfare, the warrior ethos has grounded the ethical legitimacy of violence in an assumption of personal vulnerability. This has manifested as an expectation of reciprocal inter-combatant risk.

The demands of the warrior ethos, however, have always been situated within a broader and ever-changing cultural, social, and political context. Frimer argues that heroes are those who defend the 'sacred values' of a particular group—'core values which are . . . non-negotiable' (cited in Webb 2015). Importantly though, while non-negotiable, these values can change over time to reflect a group's modifying objectives and goals (Frimer, cited in Webb 2015). The following section will explore this change—both material and ideational—in the context of war, and in the process, illustrate the shifting role of reciprocal risk within the warrior ethos. It argues that one of the most important outcomes of this shift has been the ascendancy of combat responsibility over physical risk as the primary determinant of ethical status. This analysis will commence with the early modern and modern period of Western warfare.

Early Modern and Modern Warfare

The decline of feudalism during the early modern period triggered what Sloterdijk has termed 'Hobbesian castration', the 'submission of the citizen's savage pride to the sovereignty of the state' (cited in Coker 2007: 64). In few realms was this change felt more profoundly than in war, which underwent an intensified period of instrumentalization.[12] Whereas the bearing of arms had traditionally been restricted to an empowered and volunteer warrior class, the early modern period witnessed the rise of large conscripted armies, and *levée en masse*, as well as mercenary forces. In a significant departure from the knightly combat of medieval Europe, the armies of the Napoleonic Wars consisted of a mass of poorly armed and trained recruits. The 'human wave' tactics of such armies earned them the label 'cannon fodder' (Blum 2010: 76).

Technological innovations during this and later periods included the railway, submarine, wireless communication (early radio), chemical weapons, the machine gun, increasingly powerful artillery, and rifled, breach-loading weapons. Each of these contributed to an increase in the destructive potential of war. Howard estimates that by 1914, a sole regiment of field guns could deliver to a target area more firepower in a single hour than had been discharged by all adversary powers during the Napoleonic Wars (1977: 120). This period also heralded the arrival of 'total war', an ideational shift in hostilities that saw entire populations mobilized for the military effort and a near-erasure of the targeting distinction between combatant and non-combatant. If early modern warfare can be said to have eroded notions of chivalric contest in battle, the violence of the modern age represented a fundamental challenge to its very existence.

A key element of this challenge was the reduction of individual autonomy on the battlefield, a condition amplified by the technological distancing of violence. By the early twentieth century, lethal force was routinely exchanged between adversaries who, while sharing the same physical space, never

[12] This instrumentalization was heavily influenced by the Age of Enlightenment, an intellectually, politically, and culturally formative period which functioned as a countervailing force against irrationalism, medievalism, oppression, and the subversion of democracy. Kant, one of the most prominent thinkers of the Enlightenment, viewed this period as 'mankind's exit from its self-incurred immaturity' (1954: 384). Within rationality, it was argued, lay the potential for technological and scientific advancement, material abundance, and democracy. Rationality at this time was also increasingly conceptualized in instrumental terms. Grounded in logic and scientific empiricism, instrumental rationality is orientated towards conscious calculative outcomes, and reasoned within a framework of means and ends (Brubaker 1984: 51). According to Weber, the Enlightenment had enshrined instrumental rationality in the West, translating into a modern society increasingly defined by the 'progressive [instrumental] rationalisation of all conditions of life...' (Löwith 1989: 152). Importantly, this included the institution of war. For many, including Weber, the instrumentalization of rationality was not to be celebrated, as it had led to the degeneration of core Enlightenment principles, such as individual autonomy and emancipation—trends that had particular impact on the battlefield. Horkheimer and Adorno's 'administered world' (1995), Marcuse's critique of 'technological rationality' (1991), as well as Weber's 'iron cage' (1958), sought to depict the destructive and alienating potential of reason. A similar theme was reflected in Fukuyama's classic essay, *The End of History*, in which he outlined his fear that 'courage, imagination, and idealism, [would] be replaced by economic calculation, and the endless solving of technical problems' (1989).

encountered one another face-to-face. Though long foreshadowed, industrial warfare, and particularly the two world wars, can arguably be viewed as the terminus of existential, self-empowered warriors. In their place now functioned the citizen-soldier of diminished agency. Ernst Jünger, one of the most important chroniclers of the First World War, effectively captured this loss of agentic control. Recounting the irregular artillery barrages of the frontline, Jünger wrote:

> You cower in a heap alone in a hole and feel yourself the victim of a pitiless thirst for destruction. With horror you feel that all your intelligence, your capacities, your bodily and spiritual characteristics, have become utterly meaningless and absurd. While you think it, the lump of metal that will crush you to a shapeless nothing may have started on its course... You know that not even a cock will crow when you are hit. (1929: 180)[13]

Jünger presents an image of the combatant stripped of individualism, and in the process, rendered a mere component of the greater machinery of war. In the wake of so profound a change, the traditional warrior ethos, centred on individual autonomy and valour, appeared increasingly anachronistic.

This is not to argue that physical risk concluded as a determinant of ethical legitimacy during this period; it did not. Increasingly though, when reciprocal risk was invoked, it was by those military traditionalists who objected to the speed and intensity of technological change in war. French soldier Blaise de Montluc (1502–77) was one of these traditionalists, writing the following in reference to Arquebusiers—wielders of an early muzzle-loaded firearm:

> Would to heaven that this accursed engine had never been invented... I had not then received those wounds which I now languish under, neither had so many valiant men been slain for the most part by the pitiful fellows and the greatest cowards. (1972: 41)

The Italian poet Ariosto echoed these sentiments in the early sixteenth century, describing firearms as a 'wicked and terrible discovery' that had 'destroyed martial glory, left the profession of arms without honour, and reduced valour and virtue to nought' (cited in Bell 2012). According to these accounts, firearms, by diminishing the physical risk to the bearer to an unacceptable degree, had degraded war. Indeed, opposition to this technology

[13] This quote is sourced from an earlier translation of *Storm of Steel*.

was so intense, firearm operators were often viewed more as criminals than soldiers (Howard 1977: 13).[14] Benjamin Constant reaffirmed the importance of reciprocal risk in his own work, when lamenting the loss of more direct, face-to-face encounters on the battlefield. In 1814, Constant argued that:

> [T]he new way of fighting, the changes in weapons, artillery, have deprived military life of what made it most attractive. There is no longer any struggle against danger…courage itself must be tinged with resignation or indifference…war has lost its charm. (1988: 55)

There were some who pushed back against this techno-centric view of war, arguing that the military advances of this age had yet to supplant human factors such as will, courage, and *élan*. The British Cavalry training manual of 1907 wrote that '…the rifle, effective as it is, cannot replace the effect produced by the speed of the horse, the magnetism of the charge, and the terror of cold steel' (cited in Echevarria 2007: 55). This belief in the supremacy of man over machine, typified by the enduring faith in the cavalry charge, was finally and fatally undone by the artillery and machine gun generated violence of the First World War. In June 1918, a British machine gunner wrote in his diary:

> Not far from us…were the remains of Monchy Le Preux, where the cavalry had been cut up in the early stages of the battle. My pals told me that the cavalry attack was a fiasco. It is impossible to understand the reason for throwing…cavalry against machine guns skilfully emplaced behind a screen of barbed wire. (Coppard 1969: 110)

The high-risk and long venerated charge of horse and steel had been driven into obsolescence by battlefield mechanization. French Captain (Capt.) Boucherie was among those to express a profound discomfort with these changes to the character of war. Writing in 1914, Boucherie warned that:

> Technical progress, by encouraging the hope of killing in safety and without danger, threatens to make us forget that the primary quality of a soldier is his scorn of death. (cited in Chamayou 2015: 96)[15]

[14] This aspect of the historical challenge of asymmetry will be outlined in greater detail in Chapter 5, which examines military sniping.

[15] See also, Churchill (2007: 64–5); Crozier (1937:94); and Nietzsche (1979: 131).

Inventions such as the early firearm, the machine gun, and artillery, as well as broader doctrinal and organizational shifts in war, did much to undermine the warrior ethos that had traditionally bound combatants. Yet at the same time, opposition to these very changes highlights the ongoing salience of the framework, and specifically reciprocal risk, in the ethical legitimacy of inter-combatant violence. It is also important to note that for all these changes to the character of war, the threat of violent death remained a significant and reciprocal feature of the battlefield. For as long as this remained so, an ethos of shared martial values would provide the means for soldiers to reconcile themselves to this danger. As Gen. Cardot stated in 1908: '[w]e must find a way of guiding men to their death . . . and I know that way; it lies in a spirit of self-sacrifice, and nowhere else' (1908: 89).

What we do witness with the modernization of warfare, though, is a reconstruction of this warrior code—from 'aristocratic freedom' to 'military servitude' (Walzer 2006a: 35). This will be detailed further in Chapter 4, which highlights the centrality of this notion of shared servitude—or 'victimhood'—between opposing combatants in the conventionalist account of Just War. In the specific terms of the warrior ethos, however, the erosion of combatant autonomy, in conjunction with the broader professionalization of the practice, has contributed to a gradual shift in the ethical focus of this framework. Schulzke (2016) describes this shift as being one from agent-centric virtue ethics to deontological ethics. Put simply, the emphasis within the warrior ethos changed from obligation to oneself, to obligation to others.

One outcome of this change has been a growing recognition of combat responsibility as a critical determinant of ethical legitimacy in war. This responsibility manifested in one way as an expectation of loyalty to the members of one's group—an *esprit de corps*. '[I]f courage is the domain of individual excellence', writes Kaurin, 'loyalty is the virtue of the group' (2016: 38). As individual autonomy receded on the battlefield, the importance of this virtue rose accordingly. The warrior ethos was also increasingly associated with conformity with the rules and standards of battle. Walzer summarizes this aspect of combat responsibility thusly: '[t]hough chivalry is dead and fighting unfree, professional soldiers remain sensitive (or some of them do) to those limits and restraints that distinguish their life's work from mere butchery' (2006a: 45).

The final section of this chapter will explore this in greater detail. It argues that combat responsibility has gradually asserted itself as the principal indictor of the ethical status of warriors, particularly since the emergence of 'post-heroic' warfare.

Post-Heroic Warfare

The tension between traditional interpretations of the warrior ethos and the modernization of war has culminated in what some have termed the age of 'post-heroic' warfare. According to Luttwak, who first coined the term in 1995, post-heroic warfare is motivated not by heroism and a willingness to die, but instead by an overriding desire to avoid one's own casualties (1995: 122). As Chapter 1 explored, commitment to casualty aversion intensified in the United States as a consequence of political and military defeat in Vietnam. The extent of this commitment was made clear in a 1993 US Army Field Manual, which claimed:

> The American people expect decisive victory and abhor unnecessary casualties. They prefer quick resolution of conflicts and reserve the right to reconsider their support should any of these conditions not be met.
>
> (cited in Enemark 2013a: 11)

Six years later, casualty aversion served as the principal governing framework for US military action during OAF; 'instead of a war, Kosovo was restyled a risk management strategy with missiles' (Coker 2001b: 159). At the end of the conflict, then Chairman of the Joint Chiefs of Staff, Gen. Henry Shelton and Secretary of Defense William Cohen, wrote in a joint statement: 'the paramount lesson learned from Operation Allied Force is that the well-being of our people must remain our first priority' (1999: 1). Risk exposure is framed here as not only unnecessary, but something to be avoided at virtually all cost. On the basis of this overriding goal it would seem reasonable to infer that the warrior ethos, and with it, any expectation of reciprocal risk in battle, has long since concluded.

This, however, is not the case. The martial virtues outlined within the warrior ethos continue to animate and regulate the behaviour of combatants in war. Evidence of this fact is best found not in military doctrine or political statements, but rather in the testimony of fighters themselves. The following text is an extract of a speech given by British Colonel (Col.) Tim Collins to the 1st Battalion of the Royal Irish Regiment in 2003, on the eve of battle against Iraqi forces. Within, Collins demonstrates the enduring significance of reciprocal risk in the ethics of inter-combatant violence. In addition, he reveals the extent to which this expectation is today counterbalanced by a more general commitment to combat responsibility:

> The enemy knows this moment is coming too. Some have resolved to fight and others wish to survive. Be sure to distinguish between them. There are some who are alive at this moment who will not be alive shortly. Those who do not wish to go on that journey, we will not send; as for the others I expect you to rock their world. Wipe them out if that is what they choose. But if you are ferocious in battle, remember to be magnanimous in victory... If there are casualties in war, then remember that when they got up this morning and got dressed they did not plan to die this day. Allow them dignity in death. Bury them with due reverence and properly mark their graves... If someone surrenders to you, remember that they have that right in international law, and ensure that one day they go home to their family. The ones who wish to fight... well, we aim to please. Remember, however, that if you harm your regiment or its history by over-enthusiasm in killing, or cowardice, know that it is your family that will suffer. (cited in Coady 2008a: 167–8)[16]

Though reference is made here to international law, Collins ultimately makes a far deeper appeal to his troops, centred on a shared commitment to a warrior ethos. As he further confirms, an enduring component of this ethos remains physical courage. This was made even more explicit in 2006, within the US Army's COIN field manual, which outlined 'risk taking' as 'an essential part of the Warrior Ethos' (US Army 2006: 7–21). Explicit references to courage are still made in the military codes of numerous states, including Britain, the United States, Australia, Canada, Norway, and France (Robinson 2008: 7).

It is important not to misinterpret the demands of this ethos. In no way does it imply that participants in war cannot expend maximum energy to preserve their life. What it has hitherto required, however, is the presence of physical courage, as well as a willingness to risk harm and potentially even death in the pursuit of military objectives. In terms of exposure to battlefield risk, it is helpful here to imagine a spectrum. At one extreme reside those

[16] This echoes Nietzsche, who wrote in his great work, *Thus Spake Zarathustra*: 'you may only have enemies whom you hate, not enemies you despise. You must be proud of your enemy: then the successes of your enemy are your successes too' (1979: 160).

such as the kamikaze pilot and suicide bomber, for whom death in battle is a certainty. At the other end are those who kill in the complete absence of physical risk, such as the UAV operator. Between these two extremes is the regular soldier, who seeks to kill his adversary while labouring under some measure of reciprocal vulnerability (Chamayou 2015: 84). Even by the violent standards of war, the suicide bomber is regarded as an aberration by most, a deviation from many of the core expectations those of us, particularly in the West, hold towards warfare. In reference to the Japanese kamikaze, one American Admiral wrote at the time of the 'detached horror' that comes from witnessing 'a sight so alien to western philosophy' (cited in Coker 2007: 109).[17]

Crucially though, the men and women tasked with fighting war have also traditionally recoiled from those that stray too close to the riskless end of the spectrum. The middle ground between these two extremes is exemplified in the following example. In 2002, during Operation Anaconda, one US Navy SEAL 'figured the chances of running into an enemy were probably 100 per cent' (MacPherson 2005: 4). Nevertheless, he proceeded forward: 'he weighed the risks, and they seemed reasonable. Besides, he thought at the time, this is not how we work, reducing risk to zero – otherwise, send accountants up there' (MacPherson 2005: 4).

To return to the statement of Collins, we can also witness an important shift in focus within this contemporary iteration of the warrior ethos. Alongside his appeal to physical courage is another, arguably greater demand for combat responsibility. Increasingly, this aspect of the warrior ethos is considered a more reliable measure of ethical legitimacy than mere combat risk (Blair 2012: 64). This is to be expected, for as Chapter 1 outlined, non-combatant immunity is today regarded as the most important and inviolable rule of war, a development the warrior ethos has evolved to reflect. This is not to argue that rule conformity was absent in earlier iterations of the warrior ethos. This evolution is best understood in terms of *emphasis*; specifically, the growing importance of deontological ethics. 'Moral courage' is today explicitly included in the definition of martial courage across all US military branches, as well as a number of other militaries worldwide (Kirkpatrick 2015: 210). This translates into a greater focus on the restrained use of force, and

[17] Writing in *The Washington Post*, Cohen highlighted this ongoing discomfort in relation to the modern suicide bomber: '[a]s for the Taliban fighters, they not only don't cherish life, they expend it freely in suicide bombings. It's difficult to envisage an American suicide bomber... we don't extol the bomber and parade his or her children before the TV cameras so that other children will envy them for the death of a parent. This is odd to us. This is chilling to us. This is downright repugnant' (2009). See also, Münkler (2003: 18–19).

particularly, a dedication to preserving enemy civilian life wherever possible. In order to maintain its legitimacy, contemporary Western warfare must reflect core liberal values—the principles of discrimination and proportionality must be upheld; damage to civilian infrastructure and the environment must be minimized (McCartney 2010: 414).

It is important to appreciate that despite the growing role of combat responsibility within the warrior ethos, the ethical and moral frameworks of war remain distinct. Within most of the Just War Tradition, non-combatant immunity hinges on the assumption that civilians have done nothing, or at the very least not enough, to forfeit their *right* to be free of direct and deliberate harm. Within the warrior ethos, in contrast, non-combatant immunity must be respected because to do otherwise would degrade the warrior's *own* status as an ethical agent.

One notable example of the strengthening role of combat responsibility within the warrior ethos was a suggestion made during the Afghanistan conflict to formally recognize combat troops for 'courageous restraint'. In a statement made on NATO's website in 2010, the Counterinsurgency Advisory and Assistance Team argued:

> There should be an opportunity to recognise and celebrate the troops who exhibit extraordinary courage and self-control by not using their weapons, but instead taking personal risk to de-escalate tense and potentially disastrous situations. (Freire 2010)[18]

Though this suggestion for a new medal was ultimately rejected, statements of this type are nevertheless instructive, in that they demonstrate how the contemporary warrior ethos, while retaining a respect for personal risk and courage, increasingly links these virtues to an expectation of combat responsibility.

The modernization of warfare has brought with it an intense degree of change to the warrior ethos. While this ethical framework continues to valorize the physical risk assumed by combatants, it increasingly incorporates this within a broader consideration of the conduct of those who fight.[19]

[18] This shift towards combat responsibility has not proceeded without challenge. By the summer of 2010, there was growing frustration among International Assistance Security Forces (ISAF) that the existing rules of engagement were overly restrictive. In response, Gen. David Petraeus changed the mission emphasis within Afghanistan from 'courageous restraint' to 'relentless pursuit', with the defeat of enemy forces taking precedence over concerns for collateral damage (Farrell 2017).

[19] Connell has explored this shift in the warrior ethos in relation to changing notions of masculinity. Connell connects technological shifts in war to the emergence of a 'professionalised, calculative rationality of the technical specialist' (1985: 9).

The question that logically follows is what significance does this shift in the warrior ethos have for the challenge of risk asymmetry? This is answered in Chapters 5 and 6, which examine the historical asymmetry-challenges of military sniping and manned aerial bombing respectively. It is argued that in both cases, the growing role of combat responsibility in the warrior ethos was an essential element in the gradual ethical normalization of these modes of violence.

Chapter 7 will then examine these same trends in the context of the radical asymmetry of UAV-exclusive violence. This chapter illustrates the ongoing link between the core virtues of the warrior ethos—including courage—and exposure to physical vulnerability. As such, any use of violence exercised in the total or near-total absence of personal, physical risk will struggle to attract the same degree of ethical legitimacy afforded to riskier acts in battle. For all the change undergone within the warrior ethos, it continues to assume, to a significant degree, 'the mutual availability of the bodies of combatants on both sides for injury and death' (Gusterson 2014: 202). Importantly though, the compatibility of UAV violence with the more explicit rules and standards of war suggests that this ethical tension is likely to be a surmountable one.

Conclusion

The warrior ethos functions as one of the most important and effective ways to regulate the use of violence in battle. It does this through an appeal to the ethical status of combatants themselves. The attainment and maintenance of this status has long been contingent on exposure to the physical risk of the battlefield. By this same logic, violence applied in the absence of reciprocal risk was viewed to misalign with the core expectations of the warrior ethos. *Non-risk* stands as the critical factor in the *ethical* challenge of asymmetry. Yet as this chapter further outlined, the warrior ethos is also an evolving framework that shifts to better reflect the society that sustains it. Today, combat responsibility is an increasingly dominant feature of the ethos itself, a reality that must be properly recognized when determining the significance and implications of increasingly asymmetric modes of violence.

Before this, it is important to re-emphasize that conformity with ethical principles does not mark the end of judgement in war. In order to determine the full extent of the challenge of radically asymmetric violence, it must also be weighed against the *moral* standards of the battlefield. Chapter 4 will advance this analysis through an examination of the Just War Tradition. It argues that the capability of radical asymmetry to undermine *structural* reciprocal risk calls into question, to an unprecedented degree, the *threat-capability* of those targeted. This, it is suggested, represents a challenge to the very coherence of the moral justifications for killing in war.

4
Reciprocity and Just Conduct in War

Introduction

When assessing the significance and implications of any emerging challenge to the practice of war, including those generated by asymmetric violence, analysis should extend beyond the somewhat narrow question of rule compliance. Direct violation of the explicit rules and standards of war is an obvious way in which to challenge the justifications for lethal force. Crucially, however, these same justifications may be subject to equal or greater strain when the fundamental assumptions that underpin them are called into question. In light of this potential, this book has explored the extent to which the coherence of the existing justifications for killing in war rests on *assumed* conditions of mutual threat and endangerment between antagonists.

This analysis began in the preceding chapter with an examination of the Western warrior ethos. It argued that within this ethical framework, the legitimacy of battlefield violence has long been grounded in an assumption of direct physical risk between opposing combatants. Historically, this same framework has held killing in the absence of reciprocal risk in lower regard. Yet as Chapter 3 further argued, within more contemporary iterations of the warrior ethos, combat responsibility has supplanted physical risk as the primary determinant of ethical legitimacy in war. This suggests that radically asymmetric violence, at least in terms of the warrior ethos, is a resolvable challenge. Problematically, as this chapter will next illustrate, this resolution potential is less compatible with the Just War Tradition.[1]

The Just War Tradition has historically occupied the moral middle ground between pacifism and realism. Just War firstly rejects pacifism's absolutist condemnation of war. Rather, it argues that states, as well as certain non-state actors, may have the right, and on occasion, even the moral duty to engage in armed conflict.[2] This calculation is reached through consultation with the

[1] This chapter explores the *jus in bello* component of Just War, the moral rules that apply once hostilities have commenced.

[2] John Stuart Mill summarized this view in his classic essay, 'The Contest in America': '[w]ar, in a good cause, is not the greatest evil which a nation can suffer. War is an ugly thing, but not the ugliest

Asymmetric Killing: Risk Avoidance, Just War, and the Warrior Ethos. Neil C. Renic, Oxford University Press (2020).

DOI: 10.1093/oso/9780198851462.001.0001

moral criterion of *jus ad bellum*. In addition, Just War responds to 'the universal aspiration to impose restraints on the conduct of war', even when such hostilities are deemed permissible (Clark 1988: 24). In doing so, it provides a counter to the moral scepticism of realism. Today, this commitment to battlefield restraint is embodied within the framework of *jus in bello*. These principles, by defining the subjects and objects of combat, as well as the means by which force may be justifiably exercised, establish the bounds of morally permissible violence in war. This chapter will engage the *in bello* dimensions of the Just War Tradition to determine first, the role of reciprocal risk in the moral justifications for killing in war; and second, the extent to which these justifications may be challenged within conditions of radically asymmetric violence.

This chapter is divided into three sections, each engaging a dominant account of Just War today—conventionalism, revisionism, and contractarianism. It argues that the coherence of each of these moral accounts, particularly in terms of the permissiveness of inter-combatant violence, is grounded in an assumption of reciprocal risk. Importantly though, and in contrast to the warrior ethos, it is an assumption of *inter-belligerent*, rather than inter-combatant risk. It is for this reason that the coherence of these justifications is called into question only within conditions of *radical* asymmetry, understood as military imbalance of a degree sufficient to undermine the *structural* reciprocity of war.

Conventionalism

The Just War Tradition is best understood as a rough consensus, rather than settled theory. Debate and disagreement has always occurred between differing schools of thought regarding various aspects of the doctrine. This continues today, even in relation to the most fundamental questions of Just War, including the conditions by which individuals may be morally liable for lethal targeting. The first of these competing accounts is conventionalism, long regarded as the orthodox position within the Just War Tradition. This section will begin by outlining the main theoretical claims of conventionalism, with a particular focus on the role of reciprocal risk in its justification for lethal force. As this analysis will clarify, conventionalism justifies killing in war on the basis of the material threat of the targeted. Crucially though, this

of things: the decayed and degraded state of moral and patriotic feeling which thinks nothing is worth a war, is worse' (2008: 204).

determination is reached at the structural, rather than individual level of war. This section will conclude by arguing that the coherence of this determination is undermined within conditions of radical asymmetry.

Moral Equality and Material Threat

Arguably the most influential contemporary proponent of the conventionalist position is Michael Walzer, who outlined its core tenets in his classic work *Just and Unjust Wars* (2006a). Within its pages, Walzer sought to provide moral and practical answers to some of the most fundamental questions of war, including when and against whom violence may be directed. Walzer was particularly conscious of the extent to which modernizing trends in war had depersonalized battle, and in the process, driven many chivalric practices into obsolescence. This was examined in the previous chapter, which argued that in their stead emerged a new military code, grounded in the servitude of the soldier. It is this very servitude, Walzer argues, that provides the grounds for war's moral coherence.[3] As philosopher and former soldier J. Glenn Gray articulates, '[t]he foe is a human being like yourself, the victim of forces above him over which he has no control' (1998: 159). This reflects an intuition long-held by combatants in war: that enemy soldiers, even those fighting a criminal war, are as blameless as oneself (Walzer 2006a: 36).

From this assumption, Walzer developed and defended the moral equality of combatants. This concept establishes the realm of responsibility for all fighters as equally limited to that of *jus in bello* (Walzer 2006a: 34–7).[4] Accordingly, combatants on both sides of the conflict possess identical privileges and responsibilities—rights and duties that remain independent of the just or unjust status of the conflict's overall cause. It is this *symmetry* between enemies, along with the *independence* of *in bello* from *ad bellum* responsibilities, that constitutes Walzer's 'war convention'. The convention itself, Walzer argues, is comprised of a set of norms, adopted customs, professional codes, legal guidelines, philosophical and religious instructions, and reciprocal arrangements that guide our judgements of military conduct (2006a: 44).

[3] See also, Ryan (2008: 132); and Ryan (2012: 159).

[4] For further detail on the moral equality of combatants, as well as the role of reciprocal risk within, see Coady (2008a: 155); Dinstein (2011: 492); Ignatieff (2000: 161); Johnson (1999: 124); Kahn (2002: 2); Lichtenberg (2008: 129); May (2005: 49); McMahan (2008: 21); Meisels (2007: 62); Orend (2006: 110); Rodin and Shue (2008: 3); and Wheeler (2002: 207).

As this implies, the strength of the conventionalist view is derived in part from its historical embeddedness within military thinking and practice. In reality, many of the core assumptions of conventionalism can be located within the earlier normative theory of 'Regular War'. Regular War imagines a specific type of armed conflict, waged between political communities within a weak institutional context. According to Regular War theorists, the absence of a centralized adjudication and enforcement regime, in combination with the *de jure* equality of states, hinders the allocation of moral judgement in regard to the cause of war (Kalmanovitz 2015). One of the first scholars to explore this was the Italian jurist Raphaël Fulgosius (1367–1427), who wrote: 'for how can it be known, and who is to be judge in this matter, deciding that one side wages a just war, the other an unjust war'? (2006: 229).[5] On this basis, the focus within Regular War shifts from the assigning of definitive moral judgement to the promotion of institutional mechanisms to moderate and regulate the use of armed force. Moral status on the battlefield is thus determined by the extent to which combatants adhere to mutually applicable rules of conduct (Ryan 2014: 210). As Emmerich de Vattel argued:

> [T]he rights founded on the state of war . . . in no way depend on the justice of the cause. They depend rather on the legitimacy of the means . . . on the presence of the elements constituting a war in due form.
> (cited in Reichberg 2008: 208)

It is this historical logic that Walzer draws upon in his contemporary articulation of the moral equality of combatants. With soldiers possessing equally valid claims to moral innocence, the basis for the right to kill, and just as importantly, the liability to be harmed, must be found elsewhere than direct moral culpability.

One option is to ground the permissibility of violence in the principle of duty. The oath of enlistment is, in essence, an agreement to 'subordinate . . . [one's] . . . personal judgement to that of the collective, as expressed by the orders of those appointed over them . . .' (Meyer 2012: 200–1). This emphasis on duty has significant historical purchase, influencing justifications for lethal force as far back as the fifth century.[6] Duty also remains an

[5] It was not until the eighteenth century, in the works of Christian von Wolff (1679–1754) and Emmerich de Vattel (1714–67), that this concept was developed into a normative theory of war (Kalmanovitz 2015; Reichberg 2008: 193).

[6] On the subject of private self-defence Augustine stated: 'killing others in order to defend one's own life, I do not approve of this' (cited in Christopher 2004: 41). The only situation in which

important component in the permissibility of contemporary violence. The British Military Covenant, for example, specifically states that combatants have a 'legal right and *duty* to fight and if necessary kill' (C. Taylor 2011: 2, my emphasis). Crucially, however, throughout much of history, and particularly in more recent cases of armed conflict, institutional obligations have been insufficient to override more fundamental moral commitments.[7] Chief among these is the demand that intentional injury and death be applied only to those liable to such an end. The question that remains for the conventionalist position is how best to designate such liability within conditions of moral equality?

To justify the killing of combatants in war, Walzer and others draw upon the logic of self-defence. Combatants may be targeted with lethal force on the basis of the mortal threat they pose to others. This mortal threat equates to a rights violation—the potential target has a basic claim against the aggressor to not pose an imminent threat to their life (Thomson 1991: 300–1). Yet in war, combatants are armed and specially empowered to produce this very outcome, and kill the enemy in most circumstances. They have become, it is presumed, 'dangerous men' (Walzer 2006a: 145). It is this projection of threat that renders the combatant subject to defensive killing (Kenny 1985: 10; Norman 1995: 168; Shue 2004: 111; Walzer 2006a: 145; Zehfuss 2011: 555). At the same time, it is the combatant's *exposure* to threat that justifies their own application of violence.

Augustine relaxed this stance was when the justification for violence concerned not a private individual, but rather a public good. In the context of war, Augustine argued that the moral responsibility for violence shifted to the sovereign authority. For the combatant, therefore, obedience, and by extension killing, was to be considered a duty (Augustine 2009: Book XXII: par 75). According to Augustine, 'he to whom authority is delegated [the soldier], and is but the sword in the hand of him who uses it [the sovereign], is not himself responsible for the death he deals' (1950: 27). Augustine presents the combatant as a morally disempowered agent, instrumentalized by the state in matters of war. For an elaboration of Augustine's position, see Christopher (2004: 41); Deane (1963: 143); and Eppstein (1935: 74). Gratian made this more explicit in the twelfth century: 'when a man is justly killed, it is the law, not you, who kills him' (cited in Reichberg, Syse, and Begby 2006: 119). As noted in Chapter 2, however, Augustine restricted this duty-based exculpation to the 'just' party. (Reichberg 2008: 196). For more detail on the role of duty in the rights and responsibilities of combatants in war, see Rawls (1999: 566); and Rousseau (1997: 56). For a contemporary analysis of the role of duty within the moral equality of combatants, see Mellor (2016).

[7] In perhaps the earliest recorded trial for war crimes in 1474, Peter von Hagenbach was beheaded for acts including murder, rape, and perjury, after an ad hoc tribunal rejected his defence that he was following superior orders from the Duke of Burgundy (Greppi 1999: 533–4). The limits of 'The Defence of Superior Orders' were exposed most famously during the Nuremburg Trials of 1945 and 1946. In the subsequent codification of the legal principles that underlined the prosecution of Nazi party members, Nuremburg Principle IV stated: '[t]he fact that a person acted pursuant to order of his Government or of a superior does not relieve him from responsibility under international law, provided a moral choice was in fact possible to him' (ICRC 1950).

It should be stressed here that this justification does not amount to a blanket endorsement of such violence. Self-defence is best understood as an exculpation, one that 'serves to remove or mitigate the blame attributable to an agent for the performance of a proscribed action' (Rodin 2002: 26).[8] Within these accounts, innocence is employed 'as a term of art', denoting those who 'have done nothing, and are doing nothing, that entails the loss of their rights' (Walzer 2006a: 146). 'Non-innocent', by extension, refers not to the morally guilty, but rather those who are actively engaged in harming (Anscombe 1981b: 67; Orend 2000: 111–12).[9]

As this illustrates, notions of threat and risk are fundamental to the conventionalist justification for killing in war. Enemark puts this plainly, arguing that 'exposure to physical risk is the indispensable condition for this license [to kill]' (2013a: 81). It is important, however, not to confuse the function of risk exposure in this licence with its role in the warrior ethos. Within the conventionalist context, physical risk is relevant only for as long as it indicates an incoming material threat, at which the imperilled may then direct defensive violence. *Threat*, rather than *risk*, is the essential factor in the conventionalist account of permissible killing.

By this same logic, it is the presumptive absence of reciprocal risk between aggressor and target that provides the core justification within conventionalism for non-combatant immunity. Nagel made tacit reference to reciprocal risk in his 1970s work, *War and Massacre*. Writing against the backdrop of US involvement in Vietnam, Nagel examined the factors that distinguish legitimate killing in war from criminal murder. Nagel defined murder in the context of battle as 'the deliberate killing of the harmless' (1972: 128).[10] He went on to write that combatants and non-combatants should be delineated 'on the basis of their immediate threat or harmfulness' (1972: 140).[11]

[8] See also, Zohar (2004: 742).

[9] As other authors have noted, the Latin word *innocens*, translates not into 'guiltless', but rather 'harmless' (Reichberg, Syse, and Begby 2006: 628).

[10] This reflects a long-standing view, that the immunity of non-combatants is grounded in their presumptive lack of material threat. Vattel, for example, argued that women and children '...*make no resistance*; and consequently we have no right to maltreat their persons, or use any violence against them, much less to take away their lives' (2008: 549, my emphasis). By this same standard, Vattel condemned as 'unlawful and odious...the massacre of an enemy who has surrendered and from whom we have nothing to fear...' (2008: 574). Once again, violence is prohibited against those who pose no physical risk. On similar grounds, Grotius advocated for the immunity of prisoners of war, because 'there is no danger from prisoners...' (2012: 391–2).

[11] Murphy summarizes this distinction in a similar way: '[c]ombatants are all those of whom it is reasonable to believe that they are engaged in an attempt at your destruction. Non-combatants are all those of whom it is not reasonable to believe this' (1973: 536). See also, Dill (2015b: 115); Fullinwider (1975: 93); Lazar (2015: 101, 105); and Orend (2000: 116).

Though some might wish it were otherwise, 'war is not about killing people who are morally liable to be killed; it is about killing people who may otherwise kill you' (Shue 2008: 100). To intentionally kill non-combatants is to kill beyond the scope of self-defence, and thus in violation of an individual's right to life (Fullinwider 1975: 94).

The ongoing significance of reciprocal risk within the conventionalist justification for killing was highlighted more recently in criticism directed towards a proposed reformulation of existing military rules. In a 2005 article, Israeli scholars, Kasher and Yadlin argued that in the context of counter-terrorism operations, the preservation of Israeli combatant life ought to be privileged over that of enemy non-combatants (2005: 17). This proposed re-conceptualization of battlefield discrimination was strongly criticized by Walzer and Margalit, among others, who stressed the need to maintain the critical targeting distinction between combatants and non-combatants:

> For Kasher and Yadlin, there no longer is a categorical distinction between combatants and non-combatants. But the distinction should be categorical, since its whole point is to limit wars to those – only those – who have the *capacity to injure*. (2009, my emphasis)

The justification for killing combatants—and sparing non-combatants—hinges on the threat represented by the former. It is the material threat posed by the combatant that forfeits their right to life and the reciprocal risk *between* combatants that ensures an equal right to kill.

This once again highlights the important distinction between the conventionalist conception of reciprocal risk and the type articulated within the warrior ethos. As the previous chapter outlined, the ethical status of fighters has long been contingent on the expression of virtues such as physical courage, honour, and self-sacrifice. Reciprocal risk was the necessary configuration for the realization of these virtues, and was regarded, therefore, as a desired end in itself. In contrast, the principal focus within conventionalist accounts is the status of the intended target, with the contention being that individuals render themselves liable to harm on the basis of the material threat they represent. Framed in these terms, reciprocal risk is merely the by-product of the desired end, the restriction of violence to the harmful.

This is relevant when determining the extent to which military imbalance, and specifically radically asymmetric violence, challenges the coherence of the conventionalist right to kill. For a moral challenge to emerge in this context, the asymmetry in question must pass a substantially higher threshold than the mere mitigation of inter-combatant *risk*. It must instead undermine the potential target's capacity for *threat*. This capacity, it is argued, is sufficiently imperilled only when asymmetry impacts the *structural* level of war.

Structural Risk, Status, and the Challenge of Radical Asymmetry

For conventionalist scholars, the material threat projected by combatants in war is fundamental to the invalidation of their right to life. What is less immediately clear, however, is how *imminent* this threat must be to render such individuals morally liable to defensive harm. Here, one draws more questionable lessons from history. Grotius argued that the soldier must 'spare the blood of his foes... [and]... condemn no one to death, unless to save himself from death...' (2012: 390). For killing to be justified, Grotius added, 'the danger... must be immediate and imminent in point of time' (2012: 83).[12] Pufendorf similarly argued that it is 'lawful for me to use violence against my enemy [only] till I have repulsed the danger he threaten'd me...' (1729: 837). Both statements advance a conception of imminence narrow enough to presumably render a significant portion of combatants temporarily or even permanently immune to lethal targeting. Many would rightly consider such a reading as over-stringent. In order to maintain the logic of self-defence, while avoiding such unworkable practical outcomes, the conventionalist account draws its coherence from the *structural* reciprocity of risk between belligerents.

It is first worth noting that compared to the overall size of its armed forces, the proportion of actual combat personnel in Western armies has historically been very low. Known otherwise as the 'tooth-to-tail ratio', this concept compares the proportion of Western soldiers directly involved in combat missions (tooth) to those with a supporting role (tail). As of 2015, the active-duty military end strength of the United States Air Force (USAF) was 287,000, of which only 20,300 (7 per cent) performed combat specialities (Nussbaum 2015: 76–7). Yet personnel with a non-combat function (radio-operators, military photographers, cooks, engineers and mechanics, clerks and headquarters staff, etc.), despite posing no direct and imminent threat, have always been recognized as combatants, and thus a legitimate target in

[12] There is a contradictory element to Grotius. When referencing the 'Law of Nations', for instance, he gave licence to a considerably less restrained mode of warfare (Grotius 2012: 349–52).

war. The moral coherence for this designation of targetable status rests on the structural reality of war itself, and specifically, its *collective* dimension.

Within war, 'it is a collective that defends itself against attack from another collective, rather than simply many individuals protecting their lives in a set of individual confrontations' (Zohar 1993: 615). Combatants, even those who serve a non-combat function, are trained to deliver violence, armed, and possess the right and ability to defend themselves and take part in operations if deemed necessary. They retain, in short, the *capability* to pose a lethal individual threat. More importantly though, the combatants of an army also 'contribute to the achievement of its ends' and 'in time of war they pose a *unified threat*' (Walzer 2006b: 4, my emphasis).

What this means in practical terms is that membership of the military (status) is sufficient in most cases to render combatants morally liable to harm, even those who at the time of their targeting are materially non-threatening (conduct) (Benbaji 2008: 466; Finkelstein 2012: 175; Gross 2006: 329; Meyer 2012: 187). It is this permissive logic that Walzer draws upon when writing: 'the first principle of the war convention is that, once war has begun, soldiers are subject to attack at any time (unless they are wounded or captured)' (2006a: 138). If enemy soldiers are targeted in flight, or while sleeping at their post, it is only their immediate (inter-combatant) threat that has diminished. They remain a sufficiently contributing element of the collective (inter-belligerent) threat of their side. In situations such as these, 'the "tacit contract" of "kill or be killed" is still intact, even if attenuated' (Kels 2014).[13]

This is not the case for non-combatants, whose contribution to the collective threat of their side is presumed to fall short of the degree sufficient for liability to harm. As a consequence, whereas enemy combatants may be killed as agents of their respective collective, enemy civilians must continue to be engaged through an 'individualist prism' (Statman 2012: 96–7). There are some notable exceptions to this allocation. For instance, combatants 'rendered helpless' are not to be killed, but rather treated humanely and made

[13] Zohar terms this 'extended self-defence' (2004: 735). Steinhoff also draws a useful distinction between 'narrow' interpretations of self-defence, which are ill-suited to moral judgements in war, and the more applicable 'justifying emergency theory', inspired by the German justifying emergency statute (*rechtfertigender Notstand*) (2013: 189). The latter takes into proper consideration the less immediate threats and dangers that are inherent to the battlefield.

prisoners of war (Johnson 2006: 668).[14] In these circumstances a combatant is re-individualized on the basis of their absolute defencelessness.[15]

We may rightly question whether this exemption to the targetable status of combatants goes far enough. After all, war is replete with examples of combatants who when killed, represent neither a specific individual threat, nor contribute to the overall collective threat posed by their military—at least to a meaningfully greater degree than the protected non-combatants of their side. It is crucial to recognize, however, that the moral justification for targeting combatants within the conventionalist account of Just War rests on the *presumption*, rather than certainty of liability.[16] The reality that combatants within a war zone may be subject to harm at any moment justifies the *ex ante* estimation that their adversary, no matter the circumstance, is both a specific threat, and a sufficient contributing element to the collective threat of their side. While an individualist determination of liability would be morally preferable, it cannot be practically adopted within the kill-or-be-killed circumstances of combat. Only in the most extreme and unambiguous cases of combatant defencelessness is this presumption revoked; in all others, combatant liability is collectivized.

The argument here is that this presumption, which manifests as collective, *status*-based liability, maintains its coherence only for as long as the conflict zone itself can be realistically conceptualized as a site of two-directional violence. Once asymmetry reaches a level sufficient to insulate the military of one side entirely, or near enough, from the risk of injury and death this is no longer the case. Within such conditions, the capability of the weaker side to pose a material threat has been profoundly curtailed, not merely in terms of individual encounters, but *structurally*. The scale of this mitigation of threat undermines the ability of the stronger party to draw upon even attenuated conceptions of self-defence in its justification for lethal force. This constitutes a fundamental challenge to the moral coherence of status-based targeting. Recall again the position of Walzer:

[14] The API reaffirms the principle of *hors de combat* (literally meaning 'outside out combat') first laid out in The Hague Conventions. According to the API, a combatant is *hors de combat* if he seeks to surrender; is 'in the power of an adverse party'; or has been rendered unconscious or 'otherwise incapacitated by wounds or sickness', and therefore is 'incapable of defending himself' (ICRC 1977: Art 41(2)). Combatants rendered *hors de combat* cannot be intentionally targeted with lethal force.

[15] Combatants may also be re-individualized when they commit war crimes. In these circumstances, their violence loses its moral and legal imprimatur and may become prosecutable.

[16] The need to operate under general presumptions concerning liability in war is convincingly argued elsewhere by Coady (2008a: 164).

> We are all immune to start with; our right not to be attacked is a feature of normal human relationships. That right is lost by those who bear arms '*effectively*' because they pose a danger to other people...
>
> (2006a: 145, my emphasis)

Conventionalism attaches considerable significance to the assumption that war, and the violence that constitutes it, is fundamentally distinct from the circumstances and violence of normal society. In the absence of a state of war, collectivized liability is no longer applicable and lethal force must once more be justified on a less permissive, *individualized* basis (Statman 2012: 99–100). The argument here is that while war may still technically operate within conditions of radical asymmetry, its *nature*, grounded in a structural reciprocity of risk, has been so transformed as to strain the moral premise of conventionalism. This is not to argue that the combatants of the weaker side are no longer liable to lethal targeting. What radical asymmetry calls into question is the collective *ex ante* designation of such liability.

The doctrine of the moral equality of combatants, like so much of war, is intrinsically flawed. It does not prohibit all violence, nor does it direct it only against the most deserving.[17] It instead has the more pragmatic objective of channelling the violence of war towards its participants and away from noncombatants (Coady 2008b: 127). It justifies this distinction on the basis of material threat, rather than the more difficult to establish metric of moral culpability. This has helped ensure that for all its limitations, the conventionalist approach is at least capable of preventing the worst excesses of total war. Radical asymmetry, however, risks widening the gulf between its practical salience and moral coherence to such a degree as to undermine its capability to set justifiable limits on battlefield violence.

A complete examination of the moral significance of radical asymmetry must also acknowledge the extent to which the conventionalist account has come under increasing challenge from within the Just War Tradition itself. Today, this criticism is advanced most forcefully by a growing number of 'revisionists', who propose an alternative criterion for battlefield liability, grounded in moral responsibility. The next section will examine this position in greater detail, arguing that despite its notable departure from conventionalism, many of its key claims remain grounded within an assumption of

[17] Walzer himself acknowledges this pacifist critique: '[t]he war convention...is often described as a program for the toleration of war, when what is needed is a program for its abolition. One does not abolish war by fighting it well; nor does fighting it well make it tolerable' (2006a: 45).

structural reciprocity in war. The challenge of radical asymmetry to the coherence of the revisionist account remains, therefore, a significant one.

Revisionism

The Moral Inequality of Combatants

As already outlined, Just War is less a theory than a tradition; a synthesis of various schools of thought. Given the diversity between and within these sources, it is perhaps unsurprising that the tradition itself contains a number of enduring conceptual fault lines. One of the most significant of these is the long-standing and ongoing division—roughly split into two positions—regarding how best to allocate moral responsibility on the battlefield. The first is the conventionalist position, which affirms the existence of a 'moral equality' between combatants in war, regardless of the *ad bellum* status of either side. Alongside this exists a rival conception of battlefield morality that draws a closer analogy to policing. Within a significant portion of early Christian and natural law sources, war was viewed as a judicial, unilateral instrument, with belligerent rights reserved exclusively for those with the rightful claim to a just cause (Coates 2008: 176; McMahan 2008: 19; Reichberg 2008).[18] It is this interpretation that is re-invoked by revisionist—sometimes termed neoclassical—theorists.

According to this position, a presumption of 'moral equality' between combatants fails to conform to the standards of normal morality, which regards the justice of an individual's cause as central to the permissibility of force, including self-defence. Today, this critique is led by Jeff McMahan and others,[19] who highlight that in normal society one does not render oneself liable to defensive attack simply by posing a threat to another, as this would expose those engaged in justified self-defence to permissible violence from their culpable attacker (McMahan 2008: 21).[20] This, McMahan argues, reveals

[18] This contrast between Just and Regular War did not correspond to a neat division between two distinct schools of thought. They did, however, exist as a contrasting logic, particularly from the fifteenth to the eighteenth century (Reichberg 2008: 194–5).

[19] Alongside McMahan (2005; 2009; 2010), see Rodin (2002); Fabre (2014); Frowe (2014); and Strawser (2011). For earlier accounts of this position, see Anscombe (2006); Holmes (1989); and Ramsey (1968).

[20] At various points in history, attempts have been made to give legal expression to this view. One example of this can be found in an argument put forward by the Chief British Attorney General during his closing speech at the Nuremburg Trials: '[t]he killing of combatants in war is justifiable…only when the war itself is legal. But where war is illegal…there is nothing to justify the killing, and these

the inherent flaw in Walzer's 'war convention' as a guideline for moral conduct in war. McMahan proposes an alternative criterion for liability to attack:

> A person is morally liable to attack in war by virtue of being morally responsible for a wrong that is sufficiently serious to constitute a just cause for war, or by being morally responsible for an unjust threat in the context of war. (2008: 22)

A key feature of revisionism is the belief that the morality of violence in war is analogous with the morality of individual self-defence. 'There is one morality', argues Frowe, and:

> ... everything falls under its remit. War is not a special moral sphere with special moral rules. Rather, it is a particular aspect of ordinary life to which we apply our ordinary moral rules. (2014: 123)

According to this viewpoint, it is only in terms of scale that battlefield violence and individual self-defence differ (McMahan 2009: 156). Therefore, though there may be different epistemic thresholds for defensive harm, the substantive moral principles are the same for each. A 'reductive' account of war 'reduces the rights and responsibilities of combatants entirely to the rights and responsibilities of individual persons' (Rodin 2008: 47). This shift to an individualist prism marks a significant break with the conventionalist understanding of collectivized liability, and by extension, its justification for targeting on the basis of combatant *status*.[21] For revisionists, combatant membership is not in itself sufficient to render an individual an appropriate target for lethal force. Non-combatant membership is, by this same logic, insufficient to automatically shield an individual from targeting (Steinhoff 2013: 187). This position challenges the *independence* of *in bello* restraint from *ad bellum* responsibilities, as well as the moral *symmetry* between opposing combatants in war.

For all its other implications, this shift from a material to moral-based account of liability would appear, at least on its surface, to resolve the

murders are not to be distinguished from those of any other lawless robber bands' (cited in Draper 1998: 51). The Tribunal rejected this proposal.

[21] Conventionalists respond to this criticism by claiming that revisionism fails to properly recognize the fundamental discontinuity between ordinary society and war. This leads them to mistakenly affix to moral standards applicable to the former onto the radically different context of the latter (Shue 2008: 88). 'What Jeff McMahan means to provide', argues Walzer, 'is a careful and precise account of individual responsibility in time of war. What he actually provides, I think, is a careful and precise account of what individual responsibility in war would be like if war were a peacetime activity' (2006c: 43).

potential tension between radical asymmetry and the moral justification for killing in war. The elimination of reciprocal risk between enemies, after all, is problematic only for as long as material threat itself is a relevant factor in the right to kill. Lazar suggests as much, arguing that an individual's 'vulnerability and defencelessness, if morally relevant at all…[is]…relevant only when the victim is not liable to be killed' (2015: 106).

As will next be argued, however, even within revisionism, an assumption of structural reciprocal risk remains an important factor in the moral designation of combatant liability. Just as with conventionalist accounts, therefore, the coherence of revisionism is challenged within conditions of radically asymmetric violence.

Radical Asymmetry and the Responsibility Threshold

Few aspects of the revisionist position have generated as much debate and controversy as its potential to erode the targeting distinction between non-combatants and combatants. Radical asymmetry amplifies this risk by undermining the presumption of liability among low-level combatants.

It is important to first recognize that most Just War revisionists maintain self-defence as a justifying principle for lethal force, but precondition such a claim on a finding of moral responsibility. Theories of moral-based liability hold that it is morally better to target individuals 'who are responsible for *initiating* some unjust threat than to kill the person who, at a given point in time, *poses* the threat, but bears less responsibility for it' (Statman 2012: 96, emphasis in original).[22] By this standard, some (unjust) civilians, on account of their moral responsibility for the collective military threat of their side, will

[22] Critics of revisionism point to the substantial epistemic barriers that complicate such a finding of responsibility. Even at the level of states, an accurate proportioning of *ad bellum* justice in war is frequently no easy task (O'Donovan 2003: 13; M. Schmitt 2007: 474). Over a century after the commencement of the First World War, debate is ongoing regarding how best to assign blame for its outbreak. Carl Schmitt terms this challenge of establishing a just party in war, the 'decisionist' problem (2006: 156). Gentili invoked these epistemic limits when defending 'the general principle' of an equal application of war rights (2006: 375). German philosopher Christian von Wolff also referenced this problem, when dismissing the idea that a just belligerent could pass judgement over his unjust adversary: 'since no nation can assume for itself the functions of a judge…' (cited in Reichberg 2008: 206). Problematically, these constraints are even more conspicuous at the level of individuals. This was referenced as far back as Vitoria, who wrote of the 'excusable ignorance' of combatants regarding the just or unjust status of their conflict (1991: 313). It is for this reason, conventionalists argue, that the terms 'innocence' and 'guilt' have no easy fit in calculations of lethal accountability in war (Fullinwider 1975: 93; Mavrodes 1975: 123). One potential way to reconcile this tension is the course advocated by McMahan, which is to maintain a partial *legal* equality of combatants while doing away with its moral equivalent (McMahan 2006: 47).

forfeit their immunity from direct and deliberate harm. Murphy touched upon this theme in a 1973 article titled 'The Killing of the Innocent'. Murphy contrasted:

> the octogenarian civilian in Dresden who is an avid supporter of Hitler's war effort... with that of the poor, frightened, pacifist frontline soldier who is only where he is because of duress and who intends always to fire over the heads of the enemy. (1973: 531–2)

It seems reasonable to conclude, argued Murphy, 'that the former is much more morally guilty of the war than the latter' (1973: 531–2). The conclusion that some revisionist authors might draw from this example is that the former is also a more appropriate target for lethal force. The application of this logic would bring to an end the conventional and long-standing targeting distinction between combatants and non-combatants in war. McMahan himself has admitted that a potential consequence of his position is that non-combatants may be attacked if they bear a 'high degree of responsibility for a wrong that constitutes a just cause for war' (2008: 22). The danger of this approach is that it will essentially open 'the floodgates to total war' (Lazar 2010: 188).[23]

David Rodin, one of the most influential advocates of the revisionist position, explores this in further detail, concluding that McMahan's 'asymmetry thesis' is only half correct. Rodin accepts that unjust combatants have no moral right to direct lethal force against the just. Crucially though, moral asymmetry does not grant the just combatant additional *in bello* privileges, such as the right to target enemy non-combatants (Rodin 2008: 45).[24] Non-combatants may bear considerable responsibility for an unjust war, and by extension, the unjust threats that manifest therein. However, 'meeting this threshold of fault for an unjust threat [is not in itself] sufficient for liability to defensive force' (Rodin 2008: 49). Even if we factor in the moral culpability of non-combatants, defensive force should still be restricted to those 'whose intervening action is most proximate to the threat' (Rodin 2008: 50). This is echoed by a number of other revisionist authors, who argue that only

[23] This potential was observed as early as Vattel, who warned that 'since each Nation claims to have justice on its side it will arrogate to itself all the rights of war and claim that its enemy has none'. As a consequence, 'the contest will become more cruel, more disastrous in its effect, and more difficult of termination' (Vattel 2006: 514–15). For an expansion of this argument in its contemporary context, see Clark (1988: 79). This is yet another reason why McMahan, who is sensitive to this risk, favours the maintenance of a partial legal equality of combatants.

[24] See also, Coady (2008b: 126).

individuals sufficiently 'causally connected' to unjust threats may be justifiably targeted in war (Fabre 2014: 77; Lazar 2015: 63).

Unjust combatants, at least according to Rodin, 'clearly meet the minimum threshold of responsibility to be liable for defensive force on liability accounts of self-defence' (2008: 48). For this reason, just combatants may target them without moral censure throughout the course of the war. Kershnar offers a similar interpretation, describing a combatant as:

> [A] person who is a causal and logical agent of the project to destroy his enemy or his enemy's capacity to fight. The causal condition ensures that a person is a combatant with regard to the relevant aggressive campaign; the logical condition ensures that the person has a role *closely connected* to the aggression. (2005: 130–1, my emphasis)

These same authors further contend that in contrast to combatants, non-combatants contribute to the unjust threat of their side to a degree that is insufficient for liability to harm.[25] This emphasis on casual connection/proximity demonstrates one way in which revisionist scholars can preserve the critical distinction between protected non-combatants and targetable combatants.

Yet as authors such as Lazar have highlighted, such an approach must confront what he terms the 'responsibility dilemma' (2010). This refers to the fact that in war, there exist a meaningful percentage of combatants who individually will neither cross the threshold of necessary responsibility for an unjust threat, nor represent a material threat at the time of their targeting. Take, for example, the case of the fleeing soldier. A situation not uncommon in war is that of a defending army who, in response to the fire superiority of the enemy, panic, abandon their weapons, and break in disarray. According to most revisionist accounts, killing these individuals would present no specific moral challenge. Nor would it if instead of fleeing, these same individuals were set upon while sleeping. How, one might ask, can causal connection to a threat be established in the case of these sleeping soldiers? And, if it cannot, how can their direct and deliberate targeting be justified as a necessary measure of defensive force? Recall again that the revisionist position is a 'reductive'

[25] McMahan himself partly substantiates this point, arguing that while non-combatants may be morally liable to attack, this determination 'is of course subject to necessity and proportionality conditions' (2008: 22). Elsewhere, he suggests that the overwhelming majority of non-combatants will be insufficiently responsible for an unjust threat for it to be proportionate to kill them (McMahan 2009: 225; 2011: 549).

account of war. What this means is that in contrast to conventionalism, individuals are not liable to harm on the basis of group membership, even if that group is an unjust one (Lazar 2015: 131–2; McMahan 2009: 188). Rather, to be targeted, individuals must sufficiently contribute to the collective threat of their side. How then to square this assertion with the realities of war?

It should first be acknowledged that with regard to mid and high-level members of the military, who it is reasonable to assume have a more direct role in the organization and execution of threats, establishing a causal connection will be less challenging. In the case of low-level combatants, however, the challenge is acute. Here, McMahan himself provides an effective response. Even within a 'reductive' account of war, McMahan suggests that a distinction can be drawn between normal society and the battlefield in terms of threat:

> War involves threats that consist of activities organized in phases over extended periods of time. A soldier sleeping in invaded territory has already attacked and is engaged in attacking in the same way that I am engaged in writing this essay even while I pause to make a cup of tea. (2004a: 76)

Within the boundaries of the conventional battlefield, threats may manifest at any place and time. It is a realm of *structural* danger. As McMahan writes elsewhere, within most war zones, 'conditions that would prompt [unjust combatants] to kill have a *significant probability* of arising' (2011: 548–9, my emphasis). The presumption, therefore, that all opposing unjust combatants—even those at the lowest level—sufficiently contribute to harm, and are thus liable to harm themselves, is a morally defensible one. Importantly, this holds true even in relation to those combatants who when targeted, lack at that precise moment the capacity to injure and kill. As this illustrates, revisionism, even when applying a reductive individualist lens to its determination of liability, allows for a degree of permissiveness in inter-combatant violence that is well beyond what would be acceptable in most non-war scenarios.

It is argued that the coherence of such a presumption founders at the point at which military imbalance undermines the structural reciprocity of risk that binds belligerent forces. Again, this is not to argue that the low-level combatants of the weaker side are now immune from lethal targeting. What is instead suggested is that within conditions of radical asymmetry, the *presumption* of combatant liability can no longer be credibly sustained. Instead, the moral presumption inverts—low-level combatants of the weaker side are presumptively non-liable; and remain so, unless a *sufficient* causal connection to the collective threat of their side can be conclusively established. To maintain the more permissive status quo within the context of radical asymmetry is to risk one of two outcomes. The first would be the projection of injury and death to enemy combatants on the weaker side who, as a consequence of a lack of sufficient structural reciprocity, are not fully liable to this fate. The second, and potentially more serious implication would be a lowering of the liability threshold itself, to include a greater number of 'contributing' unjust non-combatants. Such a move would place considerable strain on the existing moral limitations on violence in war.

The differences between conventionalist and revisionist accounts of Just War are significant. Crucially though, both draw upon an assumption of structural reciprocal risk, albeit to varying degrees, in their justification for killing in war. For this reason, the coherence of each is challenged by radically asymmetric violence. This is also true in relation to a third position of Just War—contractarianism.

Contractarianism

The Contractarian Right to Kill

As outlined in the opening section of this chapter, conventionalism draws a distinction between *in bello* and *ad bellum* responsibilities in war. Free from moral culpability for the cause of the conflict, the equal right of combatants to kill is grounded in their shared exposure to injury and death. Violence is justified according to the basic right of self-defence, a justification that is available to all fighters through the imposition of reciprocal risk. Like conventionalism, and in contrast to revisionism, contractarianism separates *in bello* restraint from *ad bellum* restrictions. It defends this separation, however, through a more consequentialist logic.

For contractarianism, immunity to lethal targeting is relinquished by combatants in war, not due to the individual threat they pose, but rather because both sides freely submit themselves to a fair and mutually advantageous norm (Benbaji 2008: 487). Walzer himself acknowledges this alternative conception of war, arguing that in some cases combatants will not be bound by a shared servitude, but rather 'fight freely'. Military conduct in these situations is 'governed by rules [that] rest on mutuality and consent' (Walzer 2006a: 37).

The presumption is that states possess a rational interest in preserving their capability to defend against aggression, while at the time, seek to limit the destructiveness of war. This is ensured through a shared compromise: 'combatants may be attacked with almost no restrictions, while non-combatants may not be (directly) attacked, with almost no exceptions' (Statman 2012: 97). This tacit bargain balances the demand of military necessity with the aspiration of humanitarianism. As with the conventionalist position, the *status* of combatancy is sufficient grounds for targeting. Contractarianism is distinct, however, in that it justifies this permissiveness without resorting to what many would consider an overly broad interpretation of the right to self-defence (Benbaji 2008: 487; Statman 2011: 451).

Within this account, the mutual right to kill is the outcome of a 'contract' between enemies; one shaped by informal rules and widely shared attitudes, as well as more formal treaty-based positive and customary international law (Benbaji 2011: 45). Ultimately, though, the coherence of contractarianism is centred on rational insight into best consequences. In order to properly comprehend the rational basis for the contractarian justification for violence, it is worth briefly exploring two classic works that examine war through a rationalist, contractarian lens. The first is Brandt's influential article 'Utilitarianism and the Rules of War'. The basic question Brandt asks is:

> What rules would rational, impartial people, who expected their country at some time to be at war, want to have as the authoritative rules of war –particularly with respect to the permitted targets and method of attack?
> (1972: 149–50)

Contractarianism presupposes that all relevant parties to a conflict have an interest in limited, rather than total war. A contractual relationship that renders combatants targetable and non-combatants immune represents the most efficient way to ensure this outcome. This logic is echoed in Schelling's classic work, *The Strategy of Conflict*, which employs a game theoretic approach to the subject of war. Within, Schelling explores how best to impose limitations on violence:

> To characterise the manoeuvres and action of limited war as a bargaining process is to emphasise that, in addition to the divergence of interest over the variables in dispute, there is a powerful common interest in reaching an outcome that is not enormously destructive of values to both sides. (1980: 6)

The outcome of this centuries-long bargaining process is the framework of *jus in bello*. This emphasis on military custom does not diminish the moral seriousness of these rules. According to Walzer, a principal strength of the existing war convention is the extent to which it has been internalized by those tasked with its implementation (2006a: 44–5). Even McMahan, arguably the most prominent critic of the current liability distinction between combatants and non-combatants, acknowledges the practical importance of the current rule-set that has been 'refined, tested, and adapted to the experience of war' (2004b: 731).

The formation of contractually binding limits, even those regarded as arbitrary in the deep moral sense, is crucial to restraining the worst excesses of war, including the zero-sum dynamic of total war. The question this then raises is, what impact, if any, does the emergence of unprecedented military imbalance have on the basic functionality of this contract? It is argued here that the *in bello* right to kill advanced by contractarianism is anchored to an assumption of reciprocal structural risk between warring parties. When this is absent, and the collective threat of one side nullified, the coherence of this justification for killing in war is undermined.

The One-Sided Bargain of Radical Asymmetry

In order to highlight the challenge of radical asymmetry to the contractarian position, the importance of structural reciprocity must first be established. It is self-evidently not the case that combatants in war formally waive their right against being lethally targeted by the enemy. Rather, it is assumed that the status of combatancy, as well as the privileges that go with it, 'constitutes tacit consent to the rules of an adversary institution' (Applbaum 1999: 118). Crucially though, this tacit consent between enemies holds only for as long as these rules can be realistically conceptualized as part of a 'symmetrical code', one that is at its heart 'mutually beneficial and fair' (Benbaji 2011: 45–6). A recognition of 'mutual benefit' rests on the assumption that:

> The outcome of following a symmetrical code will be better to each relevant party (i.e., individuals and decent states) – in terms of expected benefits and expected protection of rights – than the outcome of following an asymmetrical code. (Benbaji 2011: 46)

Ryan makes a similar point in *Fair and Unfair Wars*. The rules of war, Ryan argues:

> [N]eed to represent fair constraints on both parties. Indeed rules of restraint in war develop and are operable only when participants in war see themselves as engaged in a practice which includes the enemy as a participant and not merely as the focus of hostility. (2012: 169)

Reciprocity, as defined in these terms, cannot exist only in the abstract, but must be observable through 'mutually demonstrable practices' (Ryan 2014: 214). It is argued here that few aspects are as critical to the practical demonstration of this reciprocity as a mutual right, and *realistic capability*, to kill those deemed liable to harm. Chamayou endorses this view, arguing that 'the decriminalisation of warrior homicide presupposes a structure of reciprocity' (2015: 161). This reciprocal relationship of violence, risk, and threat, hinges on the acceptance by both parties 'that they may permissibly be killed in the course of war' (Hurka 2007: 210). Just as with conventionalism and revisionism, mutual threat is a key assumption in the contractarian justification for lethal force. When one side effectively withdraws its military from exposure to threat, this assumption cannot be sustained.

To be clear, this is not an assumption of equality of *outcome*—meaning an equal distribution of injury and death to all participants. What this account of Just War does assume, however, is an equality of *obligation*; a degree of fairness in the constraints imposed upon, and opportunities afforded to, both parties. Meisels reinforces this, noting that while the playing field of war may be unequal, the traditional war convention at least minimally allows for a 'fair fight' (2017: 36).

In practical terms, this translates into an expectation that both parties to a conflict make available at least some portion of their forces for targeting. For most of history, this expectation has required little elaboration, on account of the unavoidability of structural reciprocal risk in war. As already argued, though, within certain contemporary military contexts we have witnessed a fundamental shift in the *nature* of war, defined by the capability of one side to project violence in the sustained absence of reciprocal vulnerability. Within these extreme conditions of unexposed power, the functional premise of the contractarian bargain is challenged. What, asks Chamayou 'is the worth of a right to kill one another without crime when only one of the two protagonists can still enjoy the real content of the founding permission?' (2015: 162).

Just as with other moral accounts, this challenge does not derive from the capability of select individuals in battle to mitigate risk. The contractarian approach translates into a permissive framework of violence, with all combatants subject to harm, not merely those who at the time of their targeting pose an immediate and direct physical threat. Crucially though, the implicit bargain that underpins this permissiveness retains its coherence only for as long as violence remains 'a remedy open to the weak as well as the strong' (Rodin 2006: 159). When this no longer holds, and violence is monopolized by one side, the *in bello* contract is destabilized. While this does not necessarily invalidate the right of the stronger party to kill, it does call into question the moral basis for the *permissiveness* of their violence. To ignore this concern is to risk rendering the Just War Tradition 'either an ally to the powerful, or obsolete' (Killmister 2008: 122). Lauterpacht observed the following in relation to the laws of war:

> It is impossible to visualise the conduct of hostilities in which one side would be bound by rules of warfare without benefiting from them and the other side would benefit from rules of warfare without being bound by them. (2004: 549)[26]

This is as true for the implicit expectations of contractarianism as it is for the formal rules of war. Radical asymmetry functionally strips those targeted of the contractual benefits provided by the war convention, transforming them from an equal participant to 'an object of violent measures' (C. Schmitt 2006: 320). This, in turn, intensifies the sense among the weaker party that such asymmetry is not merely unfair, but a violation of the implicit provisions of the contract—a literal 'cheating' of death. The weaker party's own violations may then be framed as an attempt to merely correct a pre-existing battlefield injustice. McMahan alludes to such a risk when writing the following:

> It is in everyone's interests that such conventions be recognised and obeyed... For it is rational for each side in a conflict to adhere to them only if the other side does. Thus if one side breaches the understanding that the conventions will be followed, it may cease to be rational or morally required for the other side to persist in its adherence to them. (2004b: 730)[27]

[26] Versions of this same legal argument can be found elsewhere. Boyle and Chinkin, for example, argue in their book, *The Making of International Law*, that: '[t]he expansion of international law-making processes, and consequentially the corpus of international law, reflects a belief... that an international system subject to law benefits all international actors, the strong as well as the weak' (2007: 25). Lamp is even more explicit, claiming that for 'reciprocity to work as a motivation to observe the law, the parties to a conflict must have an approximately symmetrical potential to inflict damage upon each other' (2011: 248).

[27] This also applies legally. Anderson warns of breakdown in the laws of war when conditions emerge of 'obligation without reciprocity' (2014).

As Chapter 2 explained, no degree of radical asymmetry is sufficient to morally justify the deliberate targeting of civilians by the weaker side. Crucially though, the targeting of civilians is equally insufficient to nullify the challenge of radical asymmetry to the coherence of the contractarian bargain. Rather, what we are potentially left with is a two-directional strain on the contract—the immunity of non-combatants by the weak, and the vulnerability of combatants by the strong. In the latter case, this strain calls into question the permissiveness of inter-combatant violence.

This chapter has engaged three competing approaches to Just War—conventionalism, revisionism, and contractarianism. The justifications for lethal force advanced by each, it has been argued, are challenged by conditions of radically asymmetric violence. Within the first of these accounts, conventionalism, material threat forms the basis for liability. Revisionist scholars critique this approach, arguing that a finding of moral responsibility should precondition any threat-based determination of targetable status. This chapter has argued that despite these differences, the coherence of both positions, particularly in terms of the permissiveness of inter-combatant violence, is challenged by the emergence of radical asymmetry. This analysis concluded with an examination of contractarianism. This position is distinct in its adoption of a rationalist, rather than rights-based approach to the moral right to kill. Just as with the other accounts, however, radical asymmetry remains a challenge, specifically through its potential to call into question the underlying coherence of the adversarial bargain between belligerents in war.

Conclusion

'It is my conviction', wrote Einstein, 'that killing under the cloak of war is nothing but an act of murder' (cited in Miller 2007: 14). This statement echoes a central tenet of pacifism, which denounces war, along with the violence that constitutes it, as fundamentally illegitimate. The Just War Tradition represents a moral counter to this contention, arguing that within certain conditions combatants may indeed kill without sanction. This moral framework has long been split, however, regarding precisely what these licensing conditions are.

This chapter has examined three of the most influential accounts of Just War—conventionalism, revisionism, and contractarianism. The justification for inter-combatant violence advanced within each, it has been argued, is grounded to a significant degree in an assumption of war as a site of structural reciprocal risk between contesting belligerents. The erosion of this structural reciprocity, through conditions of radically asymmetric violence, represents a challenge to the very coherence of these accounts.

As this chapter has further clarified, there exists a significant distinction between the ethical and moral challenge of radical asymmetry. Within the warrior ethos, this challenge is grounded in the stronger party's lack of risk. In the Just War Tradition, the focus shifts to the threat-capability—or lack thereof—of the targeted. These frameworks are also distinct in terms of the degree of military imbalance necessary for this challenge to be activated. Within the warrior ethos, a mitigation of *inter-combatant* risk may be sufficient to call into question the ethical status of the combatant. This is not the case in the Just War Tradition. Rather, the coherence of the moral justifications for killing in war is challenged only within conditions of radical, *structural* asymmetry.

Understanding these differences is essential to building an accurate account of the significance and implications of the most recent and extreme iteration of battlefield asymmetry—UAV-exclusive violence. The moral tension generated by this military imbalance is qualitatively distinct from the asymmetry-challenges of the past. Chapter 5 will confirm this empirically, through an examination of the historical asymmetry-challenge of military sniping.

PART III

THE HISTORICAL AND CONTEMPORARY CHALLENGE OF ASYMMETRY

5
Military Sniping

Introduction

Despite how it may sometimes appear, particularly in moments of political crisis, present events are rarely an exact representation of the past.[1] Rather, there exist contingent particularities to all phenomena, a lesson that should caution us against the search for historical 'grand theory'. Within history one can, however, distil and draw upon useful analogy to better navigate current and emerging issues.[2] This includes the moral challenge of radical asymmetry. Radically asymmetric violence stands as only the most recent iteration of an ongoing historical phenomenon—military imbalance of a degree sufficient to challenge the existing justifications for killing in war. Problematically, the transient nature of this historical opposition has led many authors to prematurely dismiss the existence of a morally significant distinction between the risk-disparity of the past and the radical asymmetry of the present (Brooks 2012a; Gregory 2016; Hallgarth 2013: 27; Lamb 2013: 65; Strawser 2010: 356).

This book contests this dismissal, arguing that in actuality there exists a fundamental distinction, both in significance and implications, between the inter-combatant risk-disparity that framed earlier challenges and the *structural* imbalance of risk imposed within certain conditions of contemporary conflict. Radically asymmetric violence—exemplified today through the exclusive utilization of UAVs—challenges the basic assumption of war as a site of structural, inter-belligerent risk. In doing so, it threatens to undermine the coherence of the permissive moral standards of inter-combatant violence. This challenge, it is argued, represents a qualitative disjuncture from past episodes of asymmetry.

The following three chapters will confirm these findings empirically, through analysis of two historical cases and one contemporary case of

[1] The tendency by some within politics and the media to liken any bellicose leader to Hitler, and any inaction in the face of such bellicosity a redux of Chamberlain's 'appeasement', is but one example of this ahistoricism.

[2] Cousins wrote that history, if properly utilized, offers 'a vast early warning system' (1978).

Asymmetric Killing: Risk Avoidance, Just War, and the Warrior Ethos. Neil C. Renic, Oxford University Press (2020).

DOI: 10.1093/oso/9780198851462.001.0001

asymmetric violence. The first of these historical cases is the long-standing, but now resolved asymmetry-challenge of military sniping. Opposition to this practice has for most of history been grounded in the perception that it undermined the reciprocal risk between sniper and target to a degree that imperilled the ethical status of the former. With few exceptions, however, this tension did not extend to the Just War Tradition, on account of the failure of sniping to sufficiently strain the structural reciprocity of war. Analysis of the emergence, progression, and eventual resolution of the challenge of military sniping can better inform our understanding of the current phenomenon of radical asymmetry. In particular, it can help clarify the precise crossover point between morally unproblematic and problematic military imbalance in war.

The chapter first provides a historical overview of military sniping, beginning with early forms of ranged killing and concluding with the sharpshooting of the First World War. Through its utilization of distance and concealment, sniping has long provided one of the most effective means for combatants to mitigate the reciprocal risk of battle. For just as long, this asymmetric potential has drawn criticism, rooted in the perception that the sniper was over-insulated from physical vulnerability. The chapter will then clarify that in contrast to this tension with the warrior ethos, the asymmetry-challenge of sniping did not impact the Just War Tradition to a meaningful degree. The chapter concludes by examining the gradual resolution of the asymmetry-challenge of sniping. As with the tension itself, the normalization of this practice is best understood in reference to the warrior ethos, and specifically, the increasingly significant role of combat responsibility in determinations of ethically legitimate violence.

Historical Overview

First coined in the late eighteenth century by English officers serving in India, the term 'sniper' was taken from the practice of hunting 'snipe'—small, difficult to hit game birds (Pegler 2006: 15).[3] The United States Army provides the following definition for the class of sniper: '[a] soldier with special ability, training, and equipment who is designed to deliver discriminating and highly accurate rifle fire against targets... [that]... cannot be engaged successfully by the average rifle man' (1994: 1-1). Snipers have long been utilized for their

[3] Over the course of this chapter, the terms 'sniper' and 'sharpshooter' will be used interchangeably.

ability to target and kill enemies on the battlefield that would otherwise be immune from harm. At the same time, sniping has allowed the inverse—the delivery of violence in the relative absence of physical vulnerability. The following section will outline the historical development of this risk-disparity, beginning with early modes of ranged violence and concluding with the sniping of the First World War.

Early Ranged Violence

Technology has always been a major driving force of the changing character of war. The emergence and subsequent use of the atomic bomb, for example, represents an undeniable point of delineation between the practice and logic of Great Power warfare that preceded it, and that which followed. There exists no comparable critical juncture in relation to the development and impact of sniping. Rather, sniping represents the culmination of a series of incremental technological and methodological changes in ranged violence.

If sniping is understood as merely the utilization of distance, cover, and precision to kill at a diminished risk to oneself, then the practice arguably extends back to some of the earliest modes of ballistic violence, including the crossbow and longbow. An early example of this potential was provided during the Battle of Shanzhou (1004 CE), when Liao Dynasty general Xiao Talin was singled out and killed by individual crossbow fire (Whiting 2002: 322). Another antecedent of sniping was the Welsh longbow, utilized for its range and accuracy with particular effect during the Hundred Years War (Stirling 2012: 14).

More commonly, the practice of sniping is associated with the firearm, a technology that first emerged in China around the twelfth century, before gradually spreading to Europe, the Middle East, and later India (Chase 2008: 197). Due to the crudeness of these early designs, particularly in terms of accuracy and reliability, there was initial uncertainty regarding their effectiveness in battle. Up until the 1500s, there were some who still questioned whether the musket was even a credible alternative to the bow (Smythe 1590).[4] This doubt was gradually replaced, however, by a growing faith in their potential lethality. As one anonymous author wrote of the musket in 1628, 'light armours…will be of no defence to an enemy that mingles bullets with his

[4] For rebuttal, see Barwick (1592).

arrows' (A True Patriot 1974). Despite this promise, it was not the invention of the musket, but the rifle, that marked the true genesis of sniping.

The rifle first appeared in Europe in approximately 1450, and by 1515 the weapon was being regularly produced by at least one German gunmaker (Plaster 2008). It earns its name from its rifled bore—shallow spiral grooves cut inside the barrel of the weapon to impart a spin on the projectile (Rattenbury 2015). This has the effect of stabilizing the bullet in flight, a development that allowed for greater range and accuracy in combat. This battlefield advantage was famously showcased in 1643, at the siege of Litchfield Close during the English Civil War (1642–51). In the midst of the battle, parliamentary commander, Robert Greville, became one of the earliest recorded victims of sniper rifle fire when he was shot in the eye by a Royalist sniper and killed instantly (Manganiello 2004: 233).[5]

As this illustrates, one of the most enduring advantages of the sniper has been their ability to locate and eliminate high-value targets. During the American Revolutionary War (1775–83), sharpshooters were often employed specifically to target British senior officers. In one such case, during the Battle of Saratoga (1777), Pennsylvania hunter Tim Murphy was reported to have killed British Brigadier General (Brig. Gen.) Simon Fraser at an approximate range of 270m. During the Peninsular War (1808–14), an Irish rifleman exceeded this distance, killing a French General at a range somewhere between 200m and 600m (Stirling 2012: 20).[6]

On the question of asymmetry, the ability of snipers to pick out and kill otherwise secure targets, though significant in itself, tells only half the story. Of equal importance is their capacity to deliver such violence in a manner that greatly minimizes their own exposure to physical harm.

The Asymmetric Potential of Sniping

One of the most distinctive features of the practice of sniping is the extent to which it nullifies the ability of those targeted to effectively retaliate. For those

[5] By the eighteenth century, the rifle had been integrated into numerous units across European and American battlefields. During the Seven Years War (1756–63), a German battalion of over 800 men was equipped with short, heavy rifles that surpassed virtually anything on the battlefield in terms of accuracy (Pegler 2006: 41). In the same conflict, a provincial militia named 'Rogers Rangers' also employed rifled muskets against French and Native American forces (Stirling 2012: 18).

[6] To give one example of how far sniping has developed since this period, it was reported in June 2017 that a Canadian sniper had broken the world record for the longest confirmed sniper kill shot at a distance of 3540m (Fife 2017).

accustomed to more direct, face-to-face combat, this represented a profound disruption to the regular order of the battlefield. Describing an encounter with English snipers during the Napoleonic Wars (1803–15), one unnamed French officer wrote:

> I was sent out to skirmish against some of them in green, grasshoppers I call them, you call them riflemen. They were behind every bush and stone, and soon made sad havoc amongst my men, killing all the officers of my company, and wounding myself, without our being able to do them any injury. This drove me nearly to distraction. (cited in Simmons 1899: 102)

It is important not to overstate the initial impact of the sniper, particularly given the inefficiencies of early rifle designs. Col. von Heeringen highlighted these limits when writing on the use of sharpshooters during the Battle of Long Island (1776)—the first major confrontation of the American Revolutionary War:

> The riflemen were mostly spitted to the trees with bayonets. These frightful people deserve pity rather than fear. It always takes them a quarter of an hour to load, and meanwhile they were exposed to our balls and bayonets.
> (cited in The New York Times 1880: 3)

Subsequent upgrades to the rifle resolved many of these issues, such that by the time of the American Civil War (1861–5), sniping was firmly established as a deadly feature of the battlefield.[7] 'The Yankee sharpshooters are so good on a man's head', claimed William Casey of the 34th Virginia, 'I am afraid to raise my head above the ditches' (cited in Power 1998: 114–15). A Confederate lieutenant similarly observed:

> I've seen them pick a man off who was a mile away. They could hit you so far you couldn't hear the report of the gun. You wouldn't have any idea anybody was in sight of you. (Stevens 1892, cited in Wilson 1988: 36)

The pervasiveness of this threat, coupled with the frequent inability of those targeted to meaningfully respond, induced a heightened state of stress and

[7] For more detail on the technological and organizational change to sniping during this period, see Pegler (2006: 81); and J. Taylor (2011: 89).

fear that eroded morale. One North Carolina officer wrote of this during the Civil War:

> Enemy [sharpshooters] are thinning our ranks with comparative impunity – our men being compelled simply to suffer and endure – a moral effect is being produced which may prove very detrimental to our future success.
> (1864, cited in Busch 2006: 72)

In a similar account, Union Capt. John William Deforest wrote of the constant and intense dread that sniping generated. Denied any moment of true safety, men who had otherwise performed well in battle, 'broke down under the monotonous worry' of sniper fire (Deforest 1867). Innovations in concealment, including smokeless powder in the 1890s[8] and the 'ghilli' suit[9]—clothing camouflaged to resemble foliage—further enhanced the effectiveness and psychological impact of sniping.

During the First World War, both the offensive and defensive capability of sniping advanced. The upgrade of rifles from iron to optical sights allowed snipers to penetrate and roam deeper into enemy-held territory (Westwood 2005: 1098). During this same conflict the metal-armoured double loophole was developed—a small constructed opening from which the sniper could fire from concealment. Its British creator, Major (Maj.) Hesketh-Prichard described it thusly:

> [E]ven if an enemy sniper put a bullet through the front loophole, the bullet was stopped by the sliding shutter behind, unless, that is, the shot happened to be fired – a twenty to one chance – along the exact line in which one was looking through two loopholes. (1920: 71)

For all its advantages, sniping has never represented, or even promised, a truly 'riskless' mode of violence. What the practice of sniping has long enabled, though, is the imposition of a significant risk-disparity in battle, particularly in contrast to military alternatives of the period. This distortion of reciprocal risk formed the basis for the ethical opposition that emerged and developed in parallel to the practice itself.

[8] Until the introduction of smokeless powder, a muzzle flash would often expose sniper shots. To mitigate this drawback, concealment in trees was often necessary. This need to maintain a fixed position added a level of vulnerability to early sniping.

[9] The ghilli suit was utilized by the British for the first time during the Second Boer War, fought between 1899 and 1902 (Pegler 2006: 129).

Sniping and the Warrior Ethos

This section examines the historical tension between sniping and the warrior ethos. The perception that snipers killed in the absence of a sufficient degree of physical risk contributed to the view that the practice itself was an ethical transgression. Before addressing this long-standing opposition directly, this section will highlight the enduring role of the warrior ethos in the battlefield conduct of this period.

The Modernization of War and the Endurance of Chivalry

In order to gauge the impact of any destabilizing change in war, the conceptual and material conditions within which such change takes place must first be considered. This includes prevailing battlefield practices, technological norms, explicit rules and standards, as well as shared expectations and assumptions of appropriate conduct. In other words, what precisely was destabilized by the change in question? During the eighteenth century universally binding laws of war had yet to emerge. Formally established regulation, where it existed at all, most often took the form of bilateral treaties between states. Despite this, *in bello* restrictions did exist. During the American Revolutionary War, for example, Commander-in-Chief of the Continental Army, George Washington, sought to prohibit—for both ethical and pragmatic reasons—behaviour such as reprisals and denial of quarter. Responding to a British attack on the farm of a New Jersey captain, Washington stated:

> Tho' my indignation at such ungenerous conduct of the enemy might at first prompt be to retaliation, yet humanity and policy forbid measure. Experience proves, that their wanton cruelty injures rather than benefits their cause... (1932: 443)

Whatever the intentions of Washington himself, evidence suggests that practical observance of these rules was inconsistent at best. According to some studies, quarter was routinely denied to both sides throughout the American Revolutionary War (Balderston and Syrett 1975: 124).[10] In subsequent con-

[10] An important component of this relative lack of restraint was the fact that American combatants were technically in rebellion and could, therefore, be legally hanged as traitors upon capture (Selesky 1994: 76).

flicts, including the French Revolutionary Wars, Napoleonic Wars, and American Civil War, the limits of existing military regulations and informal *in bello* standards were exposed to an even greater degree by the actualities of battle.[11] Within each of these campaigns can be found the antecedents of total war—a mode of unrestricted violence in which the distinction between targetable combatant and protected non-combatant was frequently erased. Union Gen. William Sherman offered a stark summary of this emerging logic when he declared that his forces were 'not only fighting hostile armies, but a hostile people, and must make old and young, rich and poor, feel the hard hand of war' (1864).[12]

As Chapter 3 highlighted, this period of war also marked an important shift in the warrior ethos. For many, the transition from a niche warrior class to large conscripted armies had *disenchanted* warfare, stripping it of its nobility and prestige. This disillusionment was reinforced by the intensity of the technological change in war fighting that characterized this period.[13] Writing in 1888, retired Union soldier Ambrose Bierce complained that:

> [M]en of all civilised nations [were] taking a deal of pains to invent offensive weapons that will wield themselves and defences that can take the place of the human breast. A modern battle is a quarrel of skulkers trying to have all the killing done a long way from their persons. (cited in Talley 2009: 104)

Despite this somewhat bleak assessment, it would be inaccurate to regard this period as the terminus of the warrior ethos. Though tested and strained, this framework of battlefield ethics endured, as did an expectation of reciprocal risk. A particularly vivid example of this ongoing influence can be found during the Napoleonic Wars. In one encounter between English and French forces, both armies halted 400 yards from each other, allowing their videttes (mounted sentries) to engage. An English observer described the following incident:

[11] The first meaningful attempt to address this legal vacuum was the 1863 Lieber Code, which emerged during the American Civil War.

[12] Historian George M. Fredrickson described the American Civil War as the 'twilight of humanitarianism' (1968: 183–99). For more detail on the conduct of war during this period, see Farrell (2005: 76–7); Faust (2009); Johnson (1981: 281–326); and Trask (1999: 333). French Gen. Lazare Carnot—architect of the concept of *levée en masse*—offered an even earlier glimpse of the total war rationale, writing in 1793 that 'war is a violent condition...we must exterminate, exterminate to the bitter end' (cited in Reinhard 1950: 100–8). See also, Howard (1979: 1–8); and Rothenberg (1994).

[13] This reflects a broader and historically recurring distrust of the emancipatory potential of technology. See Adas (2014); Ellul (1965); Freud (1961: 38–9); Heidegger (1982: 287); and Winner (1977).

> One of their videttes, after being posted facing an English dragoon...displayed an instance of individual gallantry, in which the French, to do them justice, were seldom wanting. Waving his long straight sword, the Frenchman rode within sixty yards of our dragoon, and challenged him to single combat. We immediately expected to see our cavalry man engage his opponent, sword in hand. Instead of this, however, he unslung his carbine and fired at the Frenchman, who not a whit dismayed, shouted out, so that every one could hear him, 'Venez avec la sabre: je suis pret pour Napoleon et la belle France' [Come with the sword. I am ready for Napoleon and the beautiful France]. After having vainly endeavoured to induce the Englishmen to a personal conflict, and after having endured two or three shots from his carbine, the Frenchman rode proudly back to his ground, cheered even by our own men. We were much amused by his gallantry, while we hissed our own dragoon... (Costello 1841: 120–1)[14]

Further evidence of this chivalric influence can be located during the American Civil War. During one intense exchange, the forces of Confederate Gen. William Hardee seized heights from which they were then able to fire down upon Union forces, who, unable to move or effectively return fire, offered virtually no resistance. Though he had no obligation to cease his assault, Hardee nevertheless called it to a halt, believing that to continue in the absence of physical risk would violate his code of honour (Mills 1894: 67).[15] The role of reciprocal risk in battlefield conduct could again be witnessed in the resistance of some to emerging technologies. Mindell recounts one such example, when a crewmember of the USS *Monitor*—the first ironclad warship commissioned by the Union Navy—wondered 'whether there isn't enough danger to give us glory' (2000: 2).

Even amidst the disruptive military changes of the eighteenth and nineteenth centuries, there exists evidence for the ongoing influence of the warrior ethos. Crucially, it was precisely this framework that was most challenged by the emergence of sniping, and specifically, its mitigation of physical risk to those who practised it.

[14] For more examples of chivalric conduct during this period, see Bertaud (1988: 280–4, 296–300); Best (1986: 200–1); and Rothenberg (1994; 1997: 90–1, 113–14).

[15] In a reversal of this situation, an 1862 editorial from the *Missouri Democrat* condemned Confederate Major General (Maj. Gen.) Hindman for ordering his troops to target pickets, pilots of steamboats, and mounted officers, writing: 'who ever before heard of the commander of an army, among civilised nations, instructing his men, in a public address, to single out mounted officers in the ranks of his foe, and deliberately shoot them down? Oh! Shame upon such chivalry...' (Moore 1862: 73).

Sniping and the Chivalry Deficit

It should first be clarified that despite the changes to war outlined above, the eighteenth- and nineteenth-century battlefield remained a realm of direct, often face-to-face violence. During the American Civil War, artillery was responsible for only 5.5 per cent of wounds, in contrast to the 94 per cent of injuries sustained by rifle fire, often at close-range (Faust 2009: 41). In his account of the 1863 Battle of Champion Hill, Byers wrote that 'lines of blue and grey [faced one another] close together and fire[d] into each other's faces for an hour and a half' (1896: 449). Byers went on to praise 'the courage of the private soldiers, standing in that line of fire' (1896: 449). A similar depiction of battle was provided by Union soldier Henry Abbott who, in a letter to his father following the Battle of Gettysburg, described opposing 'rows of dead…within 15 and 20 feet apart, as near hand to hand fighting as I ever care to see' (cited in Faust 2009: 41).

Within these conditions of battle, the risk of injury and death was inescapable. Yet for those that subscribed to the warrior ethos, this risk also conferred upon their violence an ethical imprimatur that would otherwise be absent. As Chapter 3 detailed, the warrior ethos has historically held the willingness of fighters to subject themselves to the physical risk of battle as a critical component in the legitimacy of their violence. To kill in the absence of a sufficient degree of personal risk diminished the ethical status of the combatant. The capability of sniping to produce this very outcome unsurprisingly generated a tension with this framework:

> The cool calculation, the purposefulness, and the *asymmetry of risk* involved in sharpshooting rendered it even more threatening to basic principles of humanity than the frenzied excesses of heated battle.
>
> (Faust 2009: 42, my emphasis)

Though no form of incoming violence is especially welcome in war, a special disdain has long been reserved for the sniper. At its core was the perception that the sniper 'did not subject himself to the potential personal harm of battle' (J. Taylor 2011: 93). Union light artillery Private (Pvt.) Frank Wilkeson complained during the Civil War of sharpshooters who would 'sneak around trees or lurk behind stumps', and from this concealed position 'murder a few men' (cited in Gilbert 1997: 47). *The New York Herald* wrote during this same conflict of 'the fiendish hatred' held by enemy forces for Union snipers (cited in Plaster 2008). A theme consistent among accounts from this period was the *inhumanity* of the practice. 'The regular sharpshooter often seemed to me little better than a human tiger lying in wait for human blood,' wrote Union Maj. Robert Stiles (cited in Furgurson 2001: 173). Capt. Deforest was of a similar opinion, deriding the practice of sniping as a 'sickening, murderous, unnatural, uncivilised way of being' (1867). Confederate soldiers expressed a hatred for sniping of equal intensity, referring to Union sharpshooter units, who wore a green uniform for camouflage, as 'snakes in the grass' (Pindell 1993: 46).

At first glance, one might be tempted to dismiss the significance of this opposition. After all, a certain degree of hatred and scorn for the enemy is a common enough feature in most wars. An interesting aspect of this criticism, though, was the frequency with which snipers themselves shared it. During the American Revolutionary War, a young British officer named Patrick Ferguson wrote of an incident in which he spared the life of an individual who would later be identified as George Washington. Ferguson's reasoning for this was that 'it was not pleasant to fire at the back of an unoffending individual who was acquitting himself very coolly of his duty' (cited in Hibbert 2008: 157). This discomfort with the asymmetric aspect of sniping was again evident in the 1864 killing of Maj. Gen. John Sedgwick—the highest-ranking victim of sniper fire during the American Civil War. Given his status, there can be little doubt regarding the tactical benefit of this kill. Yet despite this, the man alleged to have shot Sedgwick, picket Thomas Burgess, was always reluctant to claim credit for the action. The explanation for this timidity was that 'like many other men in the 19th century... [Burgess]... regarded this method of warfare as something akin to murder' (Ray 2006: 31).[16]

Opposition to this practice can also be found among those who fought alongside, and presumably benefited from, the sniper.[17] 'I hated sharpshooters,' wrote Frank Wilkeson, 'both Confederate and Union [and] was always glad to see them killed' (1886: 120–1). Among the public, regard for the

[16] The death of Sedgwick also serves to underscore the profound military impact of sniping. When rallying his forces, Sedgwick noticed some of his men taking cover in response to the sound of bullet fire. He famously responded: '[w]hat! What! Men dodging this way for single bullets... I am ashamed of you. They couldn't hit an elephant at this distance!' Seconds later he was shot in the head by a bullet from a Whitworth rifle at an approximate range of 730m (Latimer 1998: 12).

[17] This 'benefit' was arguably more mixed in later conflicts. During the First World War, a common method of dealing with snipers was to direct artillery fire against sectors they were suspected to operate in. For the regular soldiers subjected to these barrages, the presence of sniping was predictably unwelcome. British Pvt. Huxford recalled: 'I never liked them [snipers]. They would shoot some poor Hun and bugger off then we'd get it in the neck. Our officer often used to chase them out, telling them to ply their murderous bloody trade somewhere else' (cited in Pegler 2008: 298).

practice of sniping was equally low. Wilson Homer, one of the foremost painters of the nineteenth century, was responsible for a famous Civil War illustration titled *Sharpshooter on Picket Duty*. After the war, Homer wrote to a friend on the subject of sniping:

> I looked through one of their rifles once... the... impression struck me as being as near murder as anything I could think of in connection with the army and I always had a horror of that branch of the service.
>
> (cited in Johns 2002: 35)

Even amidst the industrialized slaughter of the First World War, sniping continued to be regarded as an unacceptable deviation from the warrior ethos. Just as with earlier opposition, the basis for this was the perceived absence of a 'rough equality in lethalness' (Demetriou 2013: 305). Evidence of this can be found in the works of British Brig. Gen. F. P. Crozier. Crozier came from a long tradition of military service that had hardened him to the realities of modern battle:

> I, for my part, do what I can to alter completely the outlook, bearing and mentality of over a thousand men... The German atrocities... help to bring out the brute-like bestiality which is so necessary for victory... The British soldier is a kindly fellow... It is necessary to corrode his mentality.
>
> (1950: 42)

Crucially though, even Crozier reserved a special contempt for the practice of sniping: 'I hated it all... the game was dirty. I had to give it up. The cool, calculated murder of defenceless men was diabolical' (1937: 101). Crozier went on to argue that the 'killing by sniper is to most men – I know there are exceptions – an act of cold murder, horrible to the most callous, distasteful to all but the most perverted' (1937: 103). Emprey, in his book *First Call*, objected to sniping on similar grounds:

> To me it appeared quite all right to 'get' a man in the heat of battle, but to lie for hours and days at a time, waiting for an enemy to expose himself – then to plug him, appeared to me a little underhanded. (1918: 222)[18]

[18] For similar accounts of opposition to sniping from this period, see Brew (2004); Kwek (2015); and Vicketts (cited in Bourke 1999: 66).

Within both accounts, the nucleus of this discomfort was the absence of a sufficient degree of reciprocal risk between sniper and target, and specifically, the sniper's lack of physical risk. Tellingly, Crozier's opposition to sniping vanished in situations in which the sniper faced unambiguous peril. In reference to the sniping of oncoming tanks, Crozier suggested that 'it was different; that sense of guilt, that conscious-stricken feeling of killing a man who at that moment was not menacing you…had disappeared' (1937: 103). Guy Nightingale favoured direct bayonet fighting over 'appalling' sniping for the same reason (cited in Yeoman 2015).

For most of its history the practice of sniping has struggled to align with the warrior ethos to the same degree as other modes of battlefield violence. The principal reason for this was the widespread perception that the sniper killed in the absence of a sufficient degree of personal physical vulnerability. In contrast to this ethical tension, sniping did not represent a lasting challenge to the more permissive *moral* standards of inter-combatant violence.

Material Non-Threat and the Moral Challenge of Sniping

This section engages the moral dimensions of the long-standing asymmetry-challenge of sniping. Whereas the previous section examined this practice in relation to the ethical status of the sniper, the focus here shifts to the recipients of such violence, and specifically, the extent to which their moral liability to harm was brought into question. It is argued that sniping, on account of its failure to undermine the structural reciprocity of war, did not represent a meaningful challenge to the moral justifications for lethal force. After highlighting this, the type and degree of sanction that did apply to the practice of sniping will be outlined.

Sniping and Just Conduct in War

As previous chapters have detailed, there is an important distinction between the ethical and moral challenge of asymmetry. The principal focus of the moral frameworks of war is the *threat* projected by the target. Only those presumed to sufficiently contribute to the material threat of their side can be directly and deliberately targeted. Within the Just War Tradition itself, however, opinion has long been divided on what actually constitutes 'sufficient contribution'.

Grotius was among those who conceived such contribution in relatively strict terms, arguing that 'it is not sufficient that, by a sort of fiction, the enemy may be conceived as forming a single body' (2012: 392). Grotius went on to claim that individuals were liable to be targeted only when 'the danger' they presented was 'immediate and imminent in point of time' (2012: 83). This interpretation would considerably narrow the ambit of acceptable violence in war (Ohlin and May 2016: 72). Grotius' demand for immediacy and imminence would be difficult to reconcile with the targeting of sleeping or fleeing soldiers, or even those unimpaired, but not overtly hostile at the moment of contact.

The influence of this Grotian model of threat is evident in the few historical cases in which the practice of sniping was conceived as a moral, rather than ethical challenge. Arguably the clearest of these examples was the long-standing opprobrium attached to the sniping of sentries, pickets, and outposts in battle. These terms refer to troops that were stationed in a forward position to warn against the enemy advance. During the Napoleonic Wars, French commanders would frequently warn British pickets of impending attack. This was to prevent the British 'being killed for no purpose', a courtesy reciprocated by British forces (Glover 1974: 303). The Duke of Wellington himself remarked that 'I always encouraged this; the killing of a poor fellow of a vedette or carrying off a post could not influence the battle...' (cited in Croker 1885: 433).

During the Franco-Prussian War (1870–1), Sutherland Edward wrote that '*contrary to modern custom* the outpost duty at Strasburgh was a game of hide-and-seek, in which the forfeit of discovery was a bullet' (cited in Edmunds 1916: 10, my emphasis). Author Archibald Forbes wrote that 'in most civilised wars...the outposts of two opposing armies in ordinary circumstances did not molest each other' (1895: 152). The basis for this immunity, Forbes argued, was the lack of physical threat presented by these individuals: 'no other term than murder expresses the killing of a lone sentry by a pot shot at long range. It was like shooting a partridge sitting' (1895: 154).[19]

Though this section concerns the moral dimensions of sniping, it is worth briefly noting that opposition to this practice, at least as it concerned pickets,

[19] See also, Wilkeson (1886: 120–1). J. M. Spaight lists a number of other historical episodes in which the practice of sniping was restricted. Osman Pasha, for example, condemned sharpshooting 'as cowardly and brutal' at the 1877 Siege of Plevna during the Russo-Turkish War (cited in Spaight 1911: 104). Further evidence can be found during the 1898 Spanish–American War (Titherington, cited in Spaight 1911: 62); as well as the Second Boer War (1899–1900), during which 'Sir George White discountenanced sniping at Ladysmith' (Spaight 1911: 104).

was given legal expression in the 1863 Lieber Code, released during the American Civil War.[20] The principle of military necessity, it outlined, 'does not admit cruelty – that is, the infliction of suffering for the sake of suffering or for revenge...' (ICRC 1863: Sec I, Art 16).[21] In accordance with this, Article 69 of the Lieber Code outlawed the sniping of 'outposts, sentinels, or pickets...except to drive them in' (ICRC 1863: Sec III, Art 69). In the year before the Code's release, Lieber clarified the basis for this restriction:

> [F]iring upon single men, for no other purpose than killing them, is simple murder. There may be an object in not allowing the enemy to post a sentinel at a certain spot, but I have known warnings being given in such cases... (1862).

Returning to the moral realm, opposition to sniping during this period extended in some cases beyond the targeting of pickets to include any vulnerable opponent. Isaac Hadden wrote in 1864 of his discomfort with the use of sniping against combatants who at the time of their targeting presented no meaningful physical threat. Hadden recounted in one letter to his family the story of a fellow soldier shot in the stomach by an enemy sharpshooter while answering the call of nature. According to Hadden, it was 'nothing unusual in this country. It is worth a man's life to go to shit here' (1864: 184). This was a particular concern given the prevalence of chronic dysentery among the troops of both sides during the conflict.[22] When engaged in non-threatening activities such as sleep or defecation, soldiers of this period 'wanted to think themselves off duty as targets as well as killers' (Faust 2009: 42).

[20] According to the UK *Manual of the Law of Armed Conflict*, the Lieber Code represented 'the most important early codification of the customs and usages of war' (UK MoD 2004: 7). Among its numerous provisions, the Lieber Code established detailed protections for prisoners of war (ICRC 1863: Sec III, Art 48–80). The rules embodied in the Lieber Code were consistent with other *in bello* standards of this time, with similar manuals or codes issued by Prussia in 1870, the Netherlands in 1871, France in 1877, Russia in 1877 and 1904, Serbia in 1878, Argentina in 1881, Great Britain in 1883 and 1904, and Spain in 1893 (Green 2000: 30). Between 1870 and 1914 the laws of war entered 'its epoch of highest repute' (Best 1980: 129). Contributions to international law during this period also include The Brussels Declaration (1874); The Laws of War on Land (Oxford Manual), adopted by the Institute of International Law in 1880 (ICRC 1880); The Hague Conventions (ICRC 1907a), and the First and Second Geneva Conventions (ICRC 1864; ICRC 1906).

[21] Upon its release, some complained that a mix of realism and idealism had left the Code overly ambiguous. Confederate Secretary of War, James Seddon, wrote on 24 June 1863, that: '[a] military commander under this code may pursue a line of conduct in accordance with the principles of justice, faith, and honor, or he may justify conduct correspondent with the warfare of the barbarous hordes who overran the Roman Empire, or who, in the Middle Ages, devastated the continent of and menaced the civilization of Europe' (1863, cited in Bosco 2008).

[22] For more detail on this, see Bollet (1992).

During the First World War, the tension between the asymmetry of sniping and the moral liability of those targeted was again evident. Robert Graves, in his autobiography *Goodbye to All That*, wrote of the one instance in which he refrained from sniping a German who was neither a prisoner nor wounded:

> While sniping from a knoll in the support line...I saw a German, about seven hundred yards away...He was taking a bath in the German third line. I disliked the idea of shooting a naked man, so I handed the rifle to the sergeant with me. 'Here, take this. You're a better shot than I am.' He got him; but I had not stayed to watch. (2000: 132)

For Graves, the nakedness of the German brought their asymmetry of vulnerability into stark relief. Added to this was the private nature of the act, which, just as with the Civil War example, functioned as an implicit appeal for the maintenance of safety in good faith. The argument here, however, is that whatever the attraction of this logic, in none of these cases did the asymmetry of risk between sniper and target truly invalidate the moral liability of the latter.

It should first be noted that many of those targeted by snipers, though unable to retaliate directly, remain an imminent physical threat to others on the battlefield. In situations such as these, violence can be morally justified no matter how narrow one's conception of threat. Crucially though, even in the examples listed above—the targeting of materially non-threatening combatants—such violence is morally permissible if we dispense with the Grotian conception of threat and instead consider the structural reciprocity of risk that frames such encounters.

War is a two-level exercise. It functions firstly at the level of individuals, as a violent interaction between opposing combatants. It is at this inter-combatant level that compliance with the warrior ethos is adjudicated. War is at the same time a collective exercise; a violent contestation between belligerent forces. Within the context of the sustained and structural danger of war, activities such as bathing and sleeping represent no more than a temporary suspension of individual threat. Combatants who are killed while engaging in such activities remain morally liable to harm on account of their sufficient, ongoing contribution to the collective material threat of their side. A central claim of this book is that the coherence of this permissiveness is ultimately grounded in, and contingent upon, an assumption of war as a reciprocal exchange of *inter-belligerent* risk. This assumption functioned as the principal basis for the conformity of sniping with both conventionalist and revisionist accounts of Just War.[23]

This conformity also holds if we shift from a deontological to consequentialist conception of justified killing in war. Contractarianism is grounded in the assumption that both parties to a conflict make available some portion of their forces for targeting. This assumption remained intact within each of the examples listed above. Structural reciprocal risk ensured that the implicit adversarial bargain at the heart of the contractarian account remained a morally coherent one, regardless of the asymmetric advantage of the sniper during specific encounters.

An expectation did exist that snipers abstain from killing opposing combatants in moments of acute physical vulnerability. This abstention, however, should not be mistaken for the fulfilment of a *moral duty*. Rather, such behaviour is more accurately likened to an act of supererogation—positive action that goes beyond what is morally required. Opposition to the sniping of outposts, including its legally formalized version in Article 69 of the Lieber Code, was an expression of long-standing battlefield *custom*—grounded in the warrior ethos. In 1911, J. M. Spaight conceded that in relation to the sniping of sentries: 'there is no binding war law on the subject and the jurists are silent as to the right or wrong of the matter' (1911: 104). Spaight followed this by writing: '[i]t is certain that humane commanders will, save in very exceptional circumstances, follow the rule laid down in the... [Lieber Code]... which forbids, ordinarily, the killing of "outposts, sentinels, or pickets"' (1911: 104). Lieber himself acknowledged the ethical basis for his restriction against the sniping of military pickets, describing the rule as 'right, chivalrous, and Christian' (1862).[24]

[23] In regard to revisionism, sniping (as with all other violence) would be morally permissible only when exercised by the 'just' against the 'unjust'.

[24] Excluding Article 69, the Lieber Code departed heavily from the more restrictive, individualized and imminence-based conception of threat and liability advanced by Grotius and others. Military necessity was viewed by Lieber to convey upon the combatant a highly permissive licence to kill (Ohlin and May 2016: 97; Witt 2012: 234–6). In the words of Lieber himself: 'it is my duty to injure my enemy... the most seriously I can, in order to obtain my end... if destruction of the enemy is my object, it is not only my right, but my duty, to resort to the most destructive means' (1889: 660). A further example of this logic can be found in Lieber's acceptance of the Confederate Army's use of explosive traps in battle: 'suppose it true – what does the *jurist* think of the lawfulness of the thing? The soldier within me revolts... It seems so cowardly. The jurist within me cannot find arguments to declare it unlawful. National battles are no quixotic tournaments' (cited in Childress 1976: 59–60, emphasis in original). Elsewhere, Lieber famously dismissed the humanitarian ethics of Vattel as 'namby pamby' (cited in Johnson 1981: 189). Only five years after the release of the Lieber Code, the principle of military necessity was further codified in the 1868 St. Petersburg Declaration, which stated that: 'the only legitimate object which States should endeavour to accomplish during war is to weaken the military forces of the enemy; that for this purpose *it is sufficient to disable the greatest possible number of men*' (ICRC 1868, my emphasis).

Isaac Hadden's complaint during the American Civil War was grounded in the same ethical concern. As a number of authors have noted, to snipe at an enemy soldier while they relieved themselves violated an 'unwritten code of honour' that promoted the exercise of mercy in the face of material non-threat (Earley 2014: 182; Robertson 1988: 143; Wilkeson 1886: 120–1). Once more, it was the framework of the warrior ethos, rather than the Just War Tradition, that ultimately provided the impetus for such opposition. Similarly, during the First World War, when Graves handed his rifle to his sergeant, he did so because the situation conflicted with his own personal conception of an ethical kill. It was the buck-passing of potential guilt and unease, not of a moral infraction. Walzer shares this opinion in his own analysis of the incident, writing: 'I hesitate to say that what is involved here is a moral feeling' (2006a: 140).[25]

The absence of an explicit moral tension, however, should not mislead readers as to the significance of the opposition that did emerge to sniping. One of the clearest indicators of the salience of this challenge is the historical frequency with which snipers have been executed upon capture.

The De Facto Status of the Sniper

Though never formalized into law, those on the battlefield have often viewed the conduct of snipers as closer to that of a criminal than combatant. One explanation for this informal criminalization was the irregular status of early snipers. These un-uniformed fighters were frequently likened to rebels and outlaws, and on this basis, excluded from protections such as prisoner of war rights (Scheipers 2011). Writing in 1582, Spanish judge Balthazar Ayala summarized this logic:

> Rebels ought not to be classed as enemies, the two being quite distinct, and so it is more correct to term the armed contention with rebel subjects execution of legal process, or prosecution, and not war. (1912: 11)[26]

[25] Walzer confirms this elsewhere in an illustration of ancient tribal warfare, in which both sides agree to unfeather their arrows, in order to limit their accuracy, and thus, the destructiveness of the conflict. '[W]e can fairly condemn', argues Walzer, 'the warrior who first arms himself with the superior and forbidden weapon and hits the enemy. Yet the man he kills was liable to be killed in any case' (2006a: 42).

[26] For more detail, see Best (1980: 118); Carnahan (2012: 675); and Hull (2005: 118).

If this were the sole explanation for the treatment of snipers, then one would expect this summary 'justice' to conclude with their formal integration into standing armies. Yet the sniper faced execution long after the professionalization of the class. The principal reason for this was the asymmetry of risk between sniper and target, and specifically, the view that such violence more closely approximated murder than a legitimate act of killing. French Gen. A. É. C. J. Mortier provided an early example of this logic, when writing of his response to sniping by Austrian troops: 'to this treachery they hung up all that corps that fell into their hands, considering them not as soldiers, but assassins, and never after gave them any quarter' (cited in Thornton 1806: 52).

During the American Civil War, snipers on both sides were also frequently subjected to execution upon capture. In an 1861 article, *The New York Times* wrote that Union snipers faced 'the risk of being cut off by cavalry, or executed, as they certainly would be, if taken' (cited in Katcher 2002: 7). In response to the capture of twelve Union sharpshooters in Virginia in 1864, one Southern paper argued: 'in our estimation they are nothing but murderers creeping up on and shooting men in cold blood and should receive the fate of murderers' (cited in Faust 2009: 41–2). During the Franco-Prussian War, the execution of snipers was also influenced by the asymmetry of the violence, in addition to the irregular status of the fighters. German Crown Prince Frederick described the actions of French *franc-tireur* ('free-shooter') fighters as follows: '[s]ingle shots are fired, generally in a cunning, cowardly fashion, on patrols, so that nothing is left for us to do but to adopt retaliatory measures…' (cited in Howard 1961: 250–1).

During both the First and Second World War, the execution of captured snipers remained routine (Lauterbach 1915: 80; Page 2008; Pegler 2006: 18, 21; Pegler 2008: 296; Stafford 2007). Sniper Harry Furness wrote during the 1944 Normandy Campaign that:

> Snipers (on both sides) if captured were shot on the spot without ceremony as snipers were hated by all fighting troops; they could accept the machine gun fire, mortar and shell splinters flying around them…but they hated the thought of a sniper taking deliberate aim to kill by singling them out.
>
> (cited in National Army Museum 2015)

The informal nature of the sanction against sniping provides further evidence of the ethical basis of this challenge. Sniping violated a battlefield code and it was upon this same battlefield that punishment was meted out. What is left to explain is how a practice once likened to murder was eventually accepted as legitimate? The final section of this chapter will address the gradual resolution of this ethical tension, a process best understood, like the challenge itself, in reference to the warrior ethos.

The Limits and Resolution of the Challenge of Sniping

Today, the sniper class is a highly coveted position in most militaries. Within the United Kingdom, prospective snipers must complete a three-month training course, with a success rate of only one in four (Hegarty 2012). Research at the Military College of Canada has also found that today's snipers routinely score higher for job satisfaction and lower on tests for PTSD than average soldiers (Bradley 2010). Of even greater significance has been the shift in their ethical status. Once denigrated as cowardly and dishonourable, the modern sniper is widely regarded as an elite and discriminate professional. Two main factors explain the gradual resolution of the asymmetry-challenge of sniping. First, opposition to the practice, limited as it was to an ethical tension, was forced to give way in the face of growing military pressures, particularly from the First World War onwards. Second, the challenge itself diminished as combat responsibility was increasingly asserted as the principal determinant of ethical conduct in war.

The Exigencies of Modern War

During the First World War, opposition to sniping was particularly strong among British forces. The British, observed Gen. Lord Horne 'were slow to adopt, indeed, our souls abhorred anything unsportsmanlike' (cited in Hesketh-Prichard 1920: Foreword). As a consequence of this discomfort, the British lagged behind Germany in the implementation of sniping, including the adoption of new technologies such as telescopic sighted rifles.[27] In reference to its use for hunting, Hesketh-Prichard, an influential figure in the British sniper program, said: 'in a general way I could not help thinking that they were unsportsmanlike, as they made shooting so very easy' (1920: 3). Maintaining this notion of fairness, however, became increasingly difficult as the conflict wore on:

[27] For more detail on this early German advantage, see Hesketh-Prichard (1920: 1–5); and The Western Front Association (2008).

> [H]owever lacking in imagination, however slow to realise the importance of novel methods, once we were convinced of their necessity, once we decided to adopt them, we managed by a combination of brains and energy, pluck and endurance, not only to make up the lost ground, but to take the lead in the race. In proof of this statement I would instance Heavy Field Artillery, High Explosives, Gas, Work in the Air, etc., and many other points I could mention in which Germany started ahead of us, including Sniping Observation and Scouting.
>
> (Horne, cited in Hesketh-Prichard 1920: Foreword)

In 1916, the British Expeditionary Force formed the first officially sanctioned school for sniper training (The Western Front Association 2008).[28] A 1917 *Los Angeles Times* article detailed similar efforts by the American military, including a 'school for snipers' at Camp Kearney, an Army base near San Diego (Blackburn 1917). The formation of these schools represents an important step in the normalization of sniping.

Though military necessity played an important role in the resolution of this challenge, it was insufficient by itself to ameliorate lingering ethical discomfort for the practice. Long after the formation of these early schools, sniping remained something of a Faustian bargain for Western militaries, reluctantly utilized and then quickly discarded. At the conclusion of subsequent wars, including the Second World War and Korea, the United States military 'would distance itself from its snipers' (Staff, cited in Grossman 1995: 109–10). According to Retired US Army Lieutenant Colonel (Lt. Col.) Dave Grossman, even during the Vietnam War, 'the sniper was a tool we pulled out of the box and used, then shoved back in the box' (cited in Wood 2015). The ad hoc status of American sniping only ended in 1977, when the US Marine Corps centralized its sniper training at Quantico, Virginia. The US Army was even slower to formalize sniper training, establishing their basic sniper school at Fort Benning, Georgia in 1987 (Mast and Halberstadt 2007: 23).[29] The ethical normalization of sniping has been a more incremental process, linked to conceptual, as well as material changes in the character of war.

[28] There were initially three in total, located in Linghem, Belgium (1st Army school), Acq, France (2nd Army school), and Bouchon, France (3rd Army school) (The Western Front Association 2008).

[29] A similar pattern can be found within the British military. As late as the 1990s, the British had no formally organized sniper section, and while the standard kit of British infantry battalion's included sniper rifles, individuals rarely received sufficient training in their proper use (Page 2008: 2).

Combat Responsibility and the Sniping Ethos

Chapter 3 examined the warrior ethos in detail, arguing that an expectation of personal risk has long been a factor in the ethical permissibility of violence. As it further noted, however, the warrior ethos is not an immutable framework, but rather adaptive, responding to changes in the material and ideational character of war. A key element in the adaption of the warrior ethos has been the gradual rise of combat responsibility—a soldier's professionalism, duty to their comrades, and adherence to the rules and standards of war—as a defining component of ethical legitimacy. Though physical courage, and by extension, physical risk, remains an important element of the warrior ethos, it is increasingly viewed as a secondary consideration next to restraint and rule-conformity.

It should first be re-emphasized that the practice of sniping has never been close to risk-free. As well as the generalized threat posed by enemy combatants, snipers have long had to contend with a range of specific anti-sniping measures, including artillery strikes and eventually counter-sniper teams.[30] Pegler estimates that during the First World War, British snipers frequently had as high as a 50 per cent casualty rate (2008: 299). These figures stand as an important counterpoint to the long-standing perception of snipers as cowards.

More importantly though has been the shift in the broader ethical context within which snipers operate, and specifically, the significant increase in the volume and severity of restrictions that apply to battlefield violence. Chief among these is the moral and legal demand that violence be applied only to combatants. Civilians, in contrast, retain their immunity from direct and deliberate violence. The capability of the sniper to delineate between military and non-military targets has enhanced their ethical legitimacy. This is recognized by Parks, among others, who refers to military snipers 'as the epitome of the law of war principle of distinction' (2005: 117). Former sniper during the Vietnam War Dave Nelson offers one of the clearest accounts of this shift in status. Throughout his time in Vietnam, Nelson sought to maintain the 'concept of the warrior', but crucially, one that balanced physical courage with personal responsibility and restraint:

[30] For a primary account of these risks during the First World War, see Furniss (cited in Pegler 2006: 18).

> I could not tolerate the abuse of civilians – especially not children and women. It was a very personal thing with me…it went against everything I had been taught. That made my decision to be a sniper. Killing clean shows respect for the enemy, but to kill civilians or to lose control of yourself and your concepts in life in combat is wrong…that is respect for your enemy…that's the concept behind the warrior. Kill cleanly, kill quickly, kill efficiently, without malice or brutality. (1992: 21)

A willingness to embrace physical risk remained an important feature of this conception of the warrior ethos. Nelson went as far as to tie himself to a tree when sniping to ensure that if wounded he would be forced to fight to the death (1992: 23–4). Crucially though, in order to be worthy of esteem, risk-exposure had to be coupled with responsibility, restraint, and professional conduct.

Another element of this normalization was a growing appreciation for the capability of the sniper to alleviate the physical risk of their fellow combatants. 'Back in Vietnam…they thought we were psychopathic killers', wrote Jack Coughlin, a retired Marine sniper, 'but the whole point of our existence is to be there on overwatch to minimize the threat to our own men' (cited in Wood 2015). In reference to the 2003 Iraq War, Coughlin added: 'everybody hates snipers until you go into combat…this was a city environment, and that's like Disneyland for a sniper' (cited in Wood 2015).[31]

The 2014 film *American Sniper* offers a particularly stark example of this transformation of status. The movie tells the biographical story of former United States Navy SEAL sniper, Chris Kyle, the deadliest marksman in US military history with 160 confirmed enemy kills. Upon its release, the movie generated significant (and unintended) debate over whether or not Kyle was actually a 'hero' (Zurcher 2015). What is interesting is that this disagreement was grounded not in the legitimacy of the practice of sniping itself, but rather Kyle's own demeanour and conduct in battle. According to his critics, Chris Kyle was not worthy of veneration. His 'simplistic moral code' (O'Hehir 2015), they claimed, led him to dehumanize the enemy (Hedges 2015; Reppenhagen 2015; West 2015). Supporters rejected this analysis, arguing

[31] Interestingly, though the negative perception of snipers has significantly diminished, it has not disappeared entirely. In 2000, the Rapporteur of the Council of Europe contacted Russian federal authorities after reports of the summary execution of captured combatants 'because they were snipers' (cited in ICRC 2017). As recent as 2002, a United States Army doctrine document included denial of prisoner of war rights as a potential danger to snipers, on account of enemy frustration with their 'inability to strike back effectively' (US Army 2002).

that Kyle 'undeniably did his job [in war] better than any man who came before him' (French 2015). The asymmetry of risk so essential to the practice of sniping—and so controversial for most of its history—was not considered by either side of the debate when assessing the ethical status of Kyle.[32]

The failure of sniping to problematize the moral justifications for killing in war has limited the scope of its historical challenge to the warrior ethos. As this ethical framework has evolved and modernized, combat responsibility has become an increasingly important determinant of warrior legitimacy. The capability of sniping to be utilized in a restrained and discriminate fashion has helped resolve the ethical tension generated by its asymmetry.

Conclusion

Military sniping offers a paradigmatic example of the challenge of asymmetric violence in war. For much of its history, this practice was condemned as illegitimate, on account of the perception that the sniper killed in the absence of an ethically appropriate degree of physical risk. Yet today, this challenge has been almost entirely overcome. This is partly the outcome of the evolution and modernization of the warrior ethos itself. A more fundamental factor though, was the failure of sniping to undermine the assumption of structural reciprocity that underpins the permissive moral standards of inter-combatant violence. As a consequence of this, the moral coherence of this permissiveness remained intact.

In more recent years, however, asymmetric violence has moved significantly closer to a morally problematic crossover point. Chapter 6 will explore this in the context of the asymmetry-challenge of manned aerial bombing. Within this case can be found the antecedents of a more fundamental tension, in which military imbalance is severe enough to call into question the presumption of moral liability for the weaker side.

[32] In one notable exception, political commentator and filmmaker Michael Moore tweeted in response to the movie, 'my uncle [was] killed by [a] sniper in WW2. We were taught snipers were cowards. Will shoot u in the back[.] Snipers aren't heroes' (2015). The comment drew considerable criticism.

6
Manned Aerial Bombing

Introduction

For the challenge of asymmetry to extend beyond mere ethical discomfort, it must be of a degree sufficient to mitigate the structural, inter-belligerent risk of war. Such imbalance must be radical and entrenched across the full scope of the battlefield. The erosion of this structural reciprocity calls into question not only the compatibility of such warfare with the warrior ethos, but also more seriously, the very coherence of the permissive moral standards of inter-combatant violence. This book argues that the potential of certain emerging modes of warfare to impose such conditions marks a profound disjuncture between historical and contemporary asymmetry. This was illustrated empirically in the previous chapter, which outlined the long-standing opposition to the practice of military sniping. The basis for this historic odium was the sniper's mitigation of *inter-combatant* risk. Crucially though, the failure of sniping to go beyond this, and sufficiently undermine the structural reciprocity of the battlefield, meant that with few exceptions the asymmetry-challenge of sniping was navigated and resolved in accordance with the warrior ethos, rather than Just War theory.

The focus now shifts to manned aerial bombing, a mode of violence that has long surpassed sniping in its capability to exacerbate the disparity of risk between combatant and target. In further contrast with sniping, opposition to the practice of aerial bombing has centred for most of history on its use against civilian, rather than military targets. Given the objective of this book to determine the extent to which radically asymmetric violence challenges the existing justifications for killing *otherwise targetable* individuals in battle, it is reasonable to question the inclusion of such a case, which has historically represented a more straightforward tension with the rules and standards of war.

The utility of this case, however, is two-fold. First, like sniping, the manned aerial bombing of civilians generated a profound ethical challenge, grounded in the perception that it degraded the reciprocal risk between pilot and target to a problematic degree. An examination of this tension can further inform our understanding of the current ethical discomfort surrounding radical

Asymmetric Killing: Risk Avoidance, Just War, and the Warrior Ethos. Neil C. Renic, Oxford University Press (2020).

DOI: 10.1093/oso/9780198851462.001.0001

asymmetry. Second, it is in the context of more advanced manned aerial bombing that we obtain the first glimpse of the potential of battlefield asymmetry to destabilize not just the warrior ethos, but also the very coherence of the moral justifications for inter-combatant violence. The turning point in this historical challenge was reached in the 1990s, in the asymmetry of the First Gulf War (1991) and OAF (1999). Within both conflicts, the risk imbalance imposed by the United States was so profound, that for many observers the presumption of non-liability (grounded in a presumption of material non-threat) that had historically formed the basis for opposition to civilian bombing became increasingly applicable to the combatants of the weaker side. These episodes provide the clearest precursor to the contemporary challenge of radical asymmetry. Importantly though, and in contrast to UAV-exclusive violence, manned aerial bombing has never fully undermined the assumption of war as a site of reciprocal structural risk. The asymmetry-potential of manned aerial bombing, though profound, falls short of the moral crossover point between unproblematic and problematic killing.

The chapter first provides a historical overview of manned aerial bombing, beginning in the nineteenth century and concluding with the 1999 high-altitude bombing of OAF. Throughout this period, aerial bombing was prized for its asymmetric potential, particularly when directed against non-combatant targets. The chapter will examine, and distinguish between, the ethical and moral opposition that emerged in response to this civilian bombing. This opposition will then be reconsidered in the context of the First Gulf War and OAF. It is within these conflicts that we witness a shift in the locus of this asymmetry-challenge, from civilian to combatant targeting. Yet as will be outlined, even within the military imbalance of these episodes, structural inter-belligerent risk endured. Like sniping, therefore, the resolution of the asymmetry-challenge of manned aerial bombing is best explained in reference to the evolution of the warrior ethos, and particularly, its growing emphasis on combat responsibility over personal physical risk in determinations of ethically legitimate violence.

Historical Overview

When assessing the impact of any change in war, one must resist the temptation to fetishize the transformative potential of emerging weaponry. This is especially true of aerial bombing, which from its very inception was heralded as 'revolutionary'. The following section will provide a historical overview of air power, beginning with military ballooning and concluding with the high-altitude bombing of OAF. For much of this history, faith in the military potential of the practice far exceeded its actual performance on the battlefield. Eventually though, aerial bombing advanced to the point where its capability to impose military imbalance in war was virtually unrivalled.

Early Bombing: Military Balloons and Airships

As noted in the previous chapter, sniping did not emerge as a fully formed practice, but rather, represented the culmination of a series of incremental material changes in ranged killing. Objection to the practice developed in parallel with its increasing asymmetric potential. Theories and concepts of aviation, in contrast—both supportive and oppositional—long precede the practice itself. More than any other battlefield technology, 'the bomber was imagined before it was invented' (Sherry 1987: 1). For some early thinkers, aviation was viewed as a way to radically transform and improve politics, society, and even religion (Corn 1983). At its most extreme, this optimism translated into an almost messianic view of aviation.[1] Counterbalancing this somewhat utopian vision were those who immediately recognized the military potential of air power. Its true promise, it was argued, lay in its ability to divorce its operators from the reciprocal vulnerability long inherent to the battlefield.

The military potential of aerial flight can arguably be observed as far back as 400 BCE, in the Indian epic *Mahabharata*. The poem references Krishna Dwaipaynana Vyasa, whose enemies rode upon a winged chariot to bomb the city occupied by his followers (cited in Wragg 1986: 18). In another early example, Francesco de Lana de Terzi wrote in 1670 of a flying craft with the capability to destroy cities with 'artificial fire-works and fire-balls', from a height that ensured it could not be 'offended by those from below' (cited in Wragg 1986: 18). H. G. Wells reinforced the asymmetric potential of aerial warfare in his 1908 science fiction novel, *The War in the Air*: 'men who were neither excited, nor expect for the remotest chance of a bullet, in any danger, poured death and destruction upon home and crowds below' (1908: 84).[2]

[1] See, for example, Victor Hugo's 1864 praise of Félix Nadar's experiments with heavier-than-air flight (cited in Purdon 2010). This optimism was also present in Jules Verne's *The Clipper of the Clouds* (1891).

[2] For other nineteenth- and twentieth-century accounts, see Chanute (1894, cited in Gibbs-Smith 1970: 221); Fullerton (1894: 571–4); Tennyson (1935: 98); and Warner (cited in Gamble 1931: 39).

Predictably though, it was some time before the actual practice of aerial bombing aligned with this expectation.

One of the first serious attempts to operationalize aerial bombing on the battlefield was made during the 1793 Siege of Toulon, fought between Republican and Royalist forces in the aftermath of the French Revolution. During the battle, Joseph-Michel Montgolfier suggested a large balloon be used to drop two bombs of 'immense size' on the besieged Royal Army within Toulon. Eventual testing of the proposed apparatus, however, proved a failure and the plan was subsequently abandoned (Haydon 2000: 5).[3] Despite this modest beginning, the military application of aerial balloons remained the subject of intense interest. In 1849, aerial bombing was deployed for the first time on the field of battle, when the Austrian army used hot-air balloons to drop small bombs onto Venetian forces stationed inside Venice. The following account of the incident appeared in the *Morning Chronicle* a week after the surrender of the city:

> Two balloons armed with shrapnels ascended from the deck of the Volcano war steamer... the captain of the English brig *Frolic*, and other persons then at Venice, are said to testify to the extreme terror and the morale effect produced on the inhabitants. (cited in Holman 2009)

In his commentary of the same incident, Edmund Flagg, a onetime US consul to Venice, wrote that the besieged forces had 'so fondly, yet so mistakenly, believed [themselves] unassailable externally by any mortal foe!' (1853: 403). As these accounts highlight, the principal benefit of this early bombing was its profound psychological impact on the targeted. In contrast, its actual destructive capability was marginal. The crude design of these military balloons meant their deployment was entirely reliant on favourable winds. This flaw was eventually overcome with the invention of the airship, an innovation that dramatically increased the military effectiveness of aerial bombing.

Rudolf Martin, an early German advocate for aerial bombing, wrote in 1908 of the potential for airships to destroy British security systems, and 'soften it up' for invasion (Hastings 2013: 461). He went on to highlight the relative cost-effectiveness of the airship, eighty of which could be purchased for the price of a single dreadnought (Hastings 2013: 461). The first battlefield

[3] The following year, French forces utilized an aerial balloon for reconnaissance during their victory against a Coalition army (Great Britain, Dutch Republic, Hanover, and Habsburg Monarchy) at the Battle of Fleurus (Hallion 2003: 64).

demonstration of this capability took place on 6 August 1914, when German Zeppelins[4] dropped thirteen bombs on a Belgian fort, missing the target and killing nine civilians in the process. Even in failure the raid served to foreshadow a new era of warfare, in which air power could be drawn upon to bypass traditional defences and deliver lethal force with impunity, at least in comparison to military alternatives of the period. As the First World War progressed, Zeppelin raids intensified, with the first German aerial attack against Britain taking place on 19 January 1915. On this occasion two Zeppelins dropped bombs on King's Lynn after failing to reach Yarmouth (Burbidge 1946: 124). As both examples further highlight, though the airship increased the lethality of aerial bombing, accuracy remained an issue. In addition, the cumbersome nature of the Zeppelin made it vulnerable to ground-based incendiary fire. This second issue would be mitigated to a significant degree by the next stage of aerial bombing—the bomber aircraft. This technology signified an important advance in the asymmetric potential of aerial bombing.

The Dawn of the Aircraft and Civilian Bombing

In 1903 the Wright brothers achieved the first sustained and controlled heavier than air-powered flight. Only eight years later, during the Italo-Turkish War, Italian pilot Giulio Gavotti became the first person to successfully drop bombs from an aircraft (Johnston 2011).[5] Improvements to aircraft design in subsequent years were enough to convince many of its military promise, including British Col. Sykes of the Royal Flying Corps (RFC). In a 1913 paper on the destructive potential of aerial bombing, Sykes wrote: 'the navies of the world... have somewhat to relinquish their proud position... the air service is the foremost line' (cited in The Spectator 1913: 8).[6] One year later this confidence was tested during the First World War.

[4] The term Zeppelin refers to the most well-known pioneer of German airships, Count Ferdinand von Zeppelin. As a mark of the success of his design, this name would eventually be used to refer to all rigid airships. By 1914, the German Zeppelin was capable of unloading a payload of 2,000 pounds of bombs from an altitude of 4,250m, at a maximum speed of 136 km/h (Hylton 2014: 19).

[5] In a letter to his father, Gavotti wrote: '[t]oday I have decided to try to throw bombs from the aeroplane... It is the first time that we will try this and if I succeed, I will be really pleased to be the first person to do it' (cited in Johnston 2011).

[6] Sykes was careful not to overstate his position: 'the idea that aviation will revolutionise war is absurd' (cited in The Spectator 1913: 8).

The weaponization of aircraft took place in an ad hoc fashion during the First World War. Initially used for scouting and reconnaissance, pilots eventually armed themselves with rifles and revolvers and did battle in chaotic mid-air duels with their aerial counterparts. By 1918, these engagements had grown more sophisticated, with most fighter aircraft equipped with twin machine guns synchronized to fire through the propeller (Hacker 2006: 259). The evolution of bomber aircraft was equally rapid. By October 1914, pilots were carrying out regular bombing raids, launching grenades and other small bombs by hand at their own initiative while flying at low altitude (Burbidge 1946: 121). Unsurprisingly, this method of bombing was highly inaccurate, with pilots having considerably greater success targeting civilian infrastructure and cities, rather than specific military objects. The aerial targeting of civilians, at least initially, was arguably more the consequence of technological deficiencies in the bomber than an explicit disregard for the norms of war.

By the end of 1914, all major belligerents of the First World War, with the exception of Britain, had conducted some form of bombing raid against enemy cities. Eventually even Britain relented, launching retaliatory bombing against German cities (Hastings 2013: 462).[7] These raids served to highlight the obsolescence of existing wartime defences, and by extension, the asymmetric potential of aerial bombing. Just as with military ballooning, however, the 'asymmetry' of this period was as much a consequence of perception as outcome:

> Bombing caused such terror and a feeling of helplessness amongst even hardened ground troops, let alone ill-defended civilians... even heavy artillery fire was preferred by soldiers to bombing... The lack of adequate anti-aircraft weapons undoubtedly contributed to the feeling of helplessness, and although it was soon clear to bomber commanders that fighter escorts were necessary for the success of their missions, to those watching on the ground below, bombers... seemed almost invulnerable... (Wragg 1986: 39)[8]

In the aftermath of the First World War, faith in the potency of airpower was especially pronounced in the context of colonial violence.[9] When advising on

[7] On 1 April 1918, Britain established the Royal Air Force (RAF), the world's first independent air force. By the end of the war, the RAF had grown from less than 100 aircraft to over 22,000 (Terry 1975).

[8] For primary accounts of the dread held for aerial bombing, see Hastings (2013: 458).

[9] For a summary of technological advances in bombing during the interwar period, see van Creveld (2010: 49).

the punitive suppression of 'native rebellion', British Capt. J. B Glubb mused that the chief advantage of aerial bombing was not its potential to inflict heavy casualties:

> Their tremendous moral effect is largely due to the demoralisation engendered in the tribesman by his feeling of helplessness and his *inability to reply effectively to the attack*...hostilities against aircraft are very poor sport...tribesmen fire away their priceless ammunition with no visible effect. (1926: 778, 782, my emphasis)

For Glubb, the chief advantage of bombing was its mitigation of reciprocal risk between opponents. Sir Arthur Harris further elucidated this position in 1924, writing that 'the Arab and Kurd now know what real bombing means: [bombing] offer[s] them no real target, no opportunity for glory as warriors, no effective means of escape' (cited in Omissi 1990: 154).[10]

Within Europe itself, the interwar period was marked by a growing consensus that the true potential of aerial bombing was its utility as a terror weapon. In short, aerial bombing allowed states to bypass the boundaries (and risk) of the conventional battlefield entirely and target vulnerable cities and non-combatants (Douhet 2009: 10; Hart 1925). In 1921, Italian Gen. Giulio Douhet summarized this logic, arguing that 'to have command of the air means to be in a position to wield offensive power so great that it defies human imagination' (2009: 23). As he wrote:

> The brutal but inescapable conclusion we must draw is this: in the face of the technical developments of aviation today, in case of war the strongest army we can deploy...and the strongest navy we can dispose...will provide no effective defence... (Douhet 2009: 10)[11]

For Douhet and others, the aerial bombing of cities offered a panacea to the military immobility that had doomed European states to four years of battlefield paralysis in the trenches of the Western Front. It was further—and erroneously—believed that the mass bombing of civilians would erode popular

[10] In 1925, Hugh Trenchard, when defending the British bombing of the Middle East, wrote that 'the air is the greatest civilising influence these countries have ever known' (1925, cited in Smith 1984: 30). This links to a broader theme, regarding the comparative lack of moral restraint that has historically typified 'inter-civilizational' violence.

[11] For similar accounts of the potential of aerial bombing during this period, see Sherman (1919: 316); and Spaight (1930: 227–8).

support for the enemy war effort, and in the process, reduce the capacity and resolve of the targeted state to fight on (Biddle 2014: 27).

In the Second World War, this logic culminated in the Allied Strategic Bombing Campaign, one of the most controversial phases in the history of air power.[12] On 27–28 July 1943, a single night of British bombing against Hamburg was responsible for more civilian deaths—approximately 50,000—than all combined German air attacks directed against Britain during the war (Lindqvist 2001: Sec 202). On the night of 13–14 February 1945, Allied forces bombed the German city of Dresden with such intensity a firestorm was generated that produced temperatures of over 1800 degrees (Fahrenheit) and left, according to the most reliable numbers, between 25,000 and 35,000 dead (Biddle 2008: 424). On the night of 9–10 March of that same year, American bombers produced a similar feat in the Pacific theatre, bombing and burning to death over 100,000 residents of Tokyo (Biddle 2014: 44).[13]

From this moral nadir, the technology and doctrine of aerial bombing has moved progressively towards greater adherence to non-combatant immunity. The most significant technical contribution to this transition was provided by the Revolution in Military Affairs (RMA) of the 1990s. The RMA was originally focused on how best to secure the central front of NATO in the event of a Soviet invasion. The proposed solution was Follow-On Forces Attack (FOFA) technologies, a mix of missile and aerial attacks that could orchestrate precision strikes at a distance. This same technology, it was hoped, would grant policymakers the means to overcome 'the hoary dictum' of 'fog and friction', and impose upon the battlefield a more discriminate mode of violence (Owens and Offley 2000: 15). Through 'precision airstrikes', commanders were able to locate high value, fixed, and mobile targets, and destroy them in a manner that minimized collateral damage and friendly fire casualties (McKitrick et al. 1998).[14] The impact of this technology was equally profound

[12] For much of the conflict there was significant disagreement between the RAF and USAAF regarding bombing policy. In contrast to the night-time 'general area bombing' pursued by British forces, the United States favoured a daylight policy of 'selective bombing' against specific nodes of enemy industry (Biddle 1995). British policy rested on the conviction that daylight bombing could not be conducted effectively, combined with the belief that 'general area bombing' would better erode civilian support for the war effort. This doctrinal disagreement came to an end in 1945, when American forces relented and switched to a 'general area' policy of night-time firebombing, in both Europe and Japan (Biddle 1995).

[13] Debate continues today over all three of these incidents regarding the exact figures of civilian dead.

[14] The technology and standards of precision bombing have advanced even further today. For example, during the Second World War, bombs would, on average, land within approximately 1,200 feet of their target. In contrast, to be regarded as a precision weapon today, a 'munition must hit within three meters, or less than ten feet' of the target (cited in Dunlap 2014: 113–14).

in terms of asymmetry, with aerial bombing brought closer than ever before to a riskless mode of violence. American bombing, in particular, was increasingly characterized by both the moral imperative and technical capability to kill from a distance—'*with no or minimum casualties*' (Der Derian 2009: xxxi, emphasis in original). This new way of war was showcased in stark terms in both the First Gulf War and OAF.

From the initial use of crude military balloons, the asymmetric potential of aerial bombing has steadily advanced to a level unmatched by virtually any other mode of conventional violence. For much of its history, however, this asymmetry was measured primarily in relation to its use against civilian, rather than military targets. The next section of this chapter will examine this in greater detail, highlighting the role of reciprocal risk in the ethical and moral opposition that emerged in response to this practice.

The Asymmetry-Challenge of Civilian Bombing

This section examines the historical challenge of civilian bombing in relation to the justifications for killing in war. Within the Just War Tradition, this challenge was grounded in the conviction that non-combatants, on account of their lack of material threat, were not liable to be targeted. Before detailing this challenge, this section will highlight the extent to which civilian bombing also degraded the *ethical* status of pilots. The basis for this tension was the perception that in contrast to aerial *fighting*, civilian bombing lacked a necessary reciprocity of physical risk between pilot and target.

Aerial Bombing and the Chivalry Deficit

Aerial Fighting

Prior to the First World War, the dominant belief among military thinkers was that the next armed conflict between European powers would be short and decisive; or as former German Chancellor Bethmann Hollweg, termed it, a 'brief storm' (cited in Farrar 1972: 40). This was grounded in the assumption that technological and doctrinal shifts in war had given primacy to offensive, rather than defensive action.[15] Yet as history well documents, this 'cult of the offensive' was mistaken. Military advancements, rather than enabling swift,

[15] For more information on this logic, see Christensen and Snyder (1990); and Van Evera (1984).

decisive engagements, heralded the arrival of mechanized siege warfare on a grand scale. This was famously exemplified by the low intensity but continuous trench warfare of the Western Front. For many, the industrialized modes of killing that characterized this theatre had stripped war of its more noble qualities.

The emergence of the aircraft, and specifically, the practice of 'dogfighting'—close range aerial battle between opposing pilots—stood as one of the most powerful military counters to this trend. Among both observers and participants, aerial fighting was seen to possess an element of knightly chivalry that had been all but extinguished in the trenches of Europe. British First World War pilot Cecil Lewis echoed this sentiment, writing that aerial combat remained '[t]he only sphere in modern warfare where a man saw his adversary and faced him in mortal combat, the only sphere where there was still chivalry and honour' (cited in Robinson 2006: 155).

American pilot Bennett Molter made a similar appeal to knightly virtues, writing in 1918 that 'in many ways the fighting aviators are living much like the lives of the heroes of chivalry. Their warfare is that of man to man' (Molter 1918: 121). Former British Prime Minister David Lloyd George held the fighter pilots of the First World War in the same esteem, describing them as 'the cavalry of the clouds...the chivalry of the air' (cited in Paris 1993: 136–7).[16] Aerial warfare resurrected 'the glories of personal endeavour in a repugnant new era of industrialised slaughter' (Hastings 2013: 459). Pilots, as the instruments of this resurrection, were perceived to transcend the battlefield, not only physically, but also ethically.

A number of factors explain this specific comparison to the knightly chivalry of the medieval period. First, when pilots were initially deployed they adopted many of the duties that had for most of history been the province of cavalry, such as reconnaissance and later raiding.[17] This overlap simplified the construction of a chivalric narrative. Aiding this further was the assumption that aerial fighting, like horse cavalry, was the domain of the aristocratic classes.[18] Aerial warfare also featured a level of pageantry, symbolism, gallantry, and ritual that often drew closer comparison with medieval duels than

[16] See also, Meijering (1987: 23).

[17] Aerial reconnaissance played an important role during the First World War. For example, Allied forces were alerted to the opportunity of a counterattack at the Battle of the Marne only after a French aircraft observed the German army change direction on their approach to Paris (van Creveld 2010: 352).

[18] The importance of this image was evident in a 1942 British Air Ministry staff memorandum, which expressed concern regarding the diminishing rates of pilots from 'the middle and upper classes, who are the backbone of this country' (cited in Hastings 2013: 174).

the impersonalized ground-based violence of the period. Enemy pilots would often exchange civilities, even when actively trying to kill one another. Wing Commander F. J. Powell recounted one example of this from the First World War:

> I had a flight commander who... went up and was shot down in flames... but two days afterwards a German aeroplane came over, dropped a message bag and told us that 'the gallant Captain Pike was shot down in mortal combat and has been buried behind our lines with full military honours'. That is the sort of *ésprit d'aviator* I suppose you'd say. That spirit it held on, I think, to the very end. (cited in Lance 1976: 67)[19]

The most important factor, however, in the ethical status of aerial fighting was the nature of its violence, and specifically, the immediacy, directness, and *reciprocity* of risk between enemies. Chapter 3 argued that within the warrior ethos, a reciprocal imposition of risk has long been regarded as an essential component of ethically justified killing. Few modes of violence demonstrated greater conformity with this 'ideal type' of combat than dogfighting. The aircrew ethos encompassed virtues such as skill, duty, honour, courage, perseverance, and self-sacrifice: '[t]he common thread that connected these qualities and abilities in the eyes of the public was risk: physical danger from a combination of the enemy, the elements or the aircraft...' (Lee 2012: 7).

This risk was significant; by 1917 the life expectancy of British fighter pilots from their first combat flight was only eight days (van Creveld 2011: 28). Crucially though, it was not just the intensity of physical risk that ethically distinguished aerial fighting from other aspects of the military, but also its personalized nature. Fighter pilots delivered and received violence in a way that enhanced, rather than undermined their agency as warriors:

> You did not sit in a muddy trench while someone who had no personal enmity against you loosed off a gun, five miles away, and blew you to smithereens – and did not know he had done it! That was not fighting; it was murder. Senseless, brutal, ignoble... (Lewis 1936: 45)[20]

[19] Molter (1918: 121); and Lewis (cited in Robinson 2006: 155) offer similar personal testimony. For more detail on the chivalric aspect of this violence, see Dougherty (2013: 29–45); Paris (1993); Raleigh (1922); and Steinbeck (1942: 87). The veneration of aerial fighting could also be witnessed in a number of fictional accounts from this period. See, for example, R. Wherry Anderson's *The Romance of Air-Fighting* (1917: 24); and Hewlett's *Our Flying Men* (1917: 37).

[20] Fighter pilot Richard Hillary was critical of long-range artillery for similar reasons (cited in Bourke 1999: 48).

Lewis highlights the importance of direct, reciprocal risk in the aircrew ethos of this period. This can also be observed in the long-standing customary prohibition against the targeting of pilots parachuting from disabled aircraft.[21] Second World War P-51 pilot Richard Peterson maintained that '[n]ormally, nobody, including the Germans would shoot anybody in a parachute. It just wasn't done. I mean there's no challenge in shooting a guy in a parachute for God's sake. He's had it. You can't miss' (2007: 0:05).[22]

Glory and honour was obtained when pilots engaged materially threatening opponents. To kill in the absence of this personal risk degraded their privileged ethical status. The truth of aerial fighting was of course more complex. Like the chivalric duels of earlier periods, encounters between enemy pilots were rarely as romantic as popular myth depicted. O'Connell argues that the majority of kills achieved by successful 'Aces' during the First World War came at the expense of new, inexperienced pilots (1989: 263). This, however, did little to diminish the dominant image of aerial fighting, as a mode of violence worthy of veneration.[23]

Aerial Bombing

The practice of intensified civilian bombing emerged against the backdrop of this chivalric image. It should first be clarified that the aerial bombing of this period was obviously not without risk. The crucial ethical distinction, however, between aerial fighting and bombing hinged on the reciprocal risk—or lack thereof—that bound pilot to target. Rather than engage with an empowered and directly threatening enemy, bombing was often used against the defenceless. Such violence lacked the 'nobility' and 'chivalry' of aerial fighting and was more difficult to reconcile with the warrior ethos

[21] This was given legal expression in the (unadopted) 1923 Hague Rules of Air Warfare. According to Article 20 of these rules: '[w]hen an aircraft has been disabled, the occupants, when endeavouring to escape by means of parachute, must not be attacked in the course of their descent' (ICRC 1923). This was reiterated by Israel in its response to a 1969 ICRC Review of the Laws of War. According to the Israeli delegation, parachutists escaping from a crippled aircraft were to be treated as *hors de combat*, on account of their 'state of helplessness and military ineffectiveness' (cited in Sabel 2000: 441). Other versions of this prohibition can be found within the API (ICRC 1977: Art 42), as well as the United States Manual: FM 27–2 (US Army 1984), and the United Kingdom Manual of the Law of Armed Conflict (UK MoD 2004: 58). The right of escaping parachutists to be free of direct and deliberate violence has today gained almost unanimous support in legal manuals and textbooks.

[22] Sandoz, Swinarski, and Zimmermann reflect upon this sentiment in the API Commentary: 'the adversary who had been brought down in flame was entitled, not to bullets, but to a salute as he went down, to wishes for his recovery if he were wounded, and flowers if he were dead' (1987: 494).

[23] It is also important to note the discomfort pilots often felt when their violence misaligned with their expectation of honourable combat. British Ace pilot James McCudden claimed that 'I hate to shoot the Hun down without him seeing me, for although this method is in accordance with my doctrine, it is against what little sporting instincts I have left' (2009: 253).

(Meilinger 1993: 116). Even before the emergence of the aircraft, evidence of this disdain can be found. On 18 January 1799, the Napoleon-led French government disbanded the *Corps d'Aerostiers*, effectively ending their experiment with military ballooning. The reason for this was the questionable practical value of the technology, combined with Napoleon's perception that it 'was not a gentlemanly act' (Christopher 1956: 27).

A similar ethical discomfort can be found in relation to the civilian bombing of the two world wars. RAF pilot Hubert R. Allen, who himself fought in the Battle of Britain, wrote:

> I was a fighter pilot, never a bomber pilot, and I thank God for that. I do not believe I could ever have obeyed orders as a bomber pilot; it would have given me no sense of achievement to drop bombs on German cities. (1972: ix)

Allen makes no appeal to the inherent rights of the targeted in this statement. Rather, his preference for aerial fighting over bombing derives from the greater conformity of the former with the warrior ethos, and specifically, inter-combatant risk. Further evidence of the ethical distinction between aerial fighting and bombing can be found in the work of Brig. Gen. F. P. Crozier. In 1937, Crozier wrote of his admiration for the pilot:

> Our British airman is a brave fellow, and there is none finer. He is prepared to be shot up or killed when the time comes – as most British servicemen are prepared...He is chivalrous and unselfish and any day would jump into the sea or in front of a train or into a burning building to risk his own life in order to save the life of a foreign woman or child. (1937: 104–5)

Crozier went on to illustrate that when these same pilots were ordered to target the defenceless, their ethical status was undermined:

> One wonders what...the thoughts of the first German airman [were] as he let loose his cargo of death and mutilation over the homes of innocent non-combatants. Was he able to calm his conscience by saying to himself: 'this is not murder; this is war. This is not *me*; this is a soldier doing his duty, obeying the orders of those in command'. (1937: 104, emphasis in original)

Ethical opposition to aerial bombing could also be found among policy-makers. British War Minister, Lord Derby, drew upon the tenets of the warrior ethos in his criticism of the proposal to bomb German cities during the

First World War: 'it would be better to be defeated, retaining honour, chivalry and humanity, rather than obtain a victory by methods which have brought upon Germany universal execration' (cited in Hylton 2014: 23).

Civilian bombing, like military sniping, mitigated the reciprocal risk between combatant and target to a degree that imperilled the ethical status of the former. This chapter will next examine the *moral* dimensions of this asymmetry-challenge. It argues that reciprocal risk remained an important component in the moral opposition to civilian bombing. Crucially though, the focus of this opposition shifted from the ethical status of the pilot, to the moral (non-)liability of the target.

Material Non-Threat and the Moral Challenge of Civilian Bombing

The first attempt to explicitly prohibit aerial bombing took place in 1899, when delegates from twenty-six states met at The Hague with the intention of imposing limits on the destructiveness of war. Among the issues discussed was the military practice of dropping bombs and other projectiles from balloons. The US delegate, Capt. William Crozier, sought to prohibit this tactic, on the basis that bombs dropped in this fashion could not be directed against a specific location. Instead, Crozier feared they 'would fall, like useless hailstones, on combatants and non-combatants alike' (cited in Kennett 1991: 2). Crozier was careful, however, to only advocate for a temporary ban on aerial bombing, his belief being that within five years a more accurate steerable balloon would be available (Ziegler 1994: 753–4). The agreement that resulted from the conference prohibited 'for a term of five years, the discharge of projectiles and explosives from balloons or by other new methods of a similar nature' (ICRC 1907b). This was replaced by a similar declaration at the Second Hague Convention of 1907, with the intention being that it would be extended to the conclusion of a Third Peace Conference in 1914. The outbreak of the First World War, however, ended hopes of a third conference, and with it, the chance for an enforceable prohibition on aerial bombing.[24]

[24] Application of the 1907 declaration was conditioned by a 'general participation' clause—essentially rendering it ineffective if not ratified by all belligerents in the relevant conflict. As a result of this, the early prohibition against aerial bombing was binding to none (Terry 1975). For more detail on the state of the law of aerial bombing, both before and after the First World War, see Biddle (1994: 142–4); Cole and Cheesman (1984: 6); Evangelista (2014: 9); Garraway (2014: 92); ICRC (1938); and Terry (1975).

Despite this setback, many remained confident that bombing would not be used against civilian targets. Director-General of Military Aeronautics at the British War Office and pilot, David Henderson, argued in 1914 that the bombing of undefended towns would never take place. His reasoning was that 'no enemy would risk the odium such action would involve' (cited in Vanderbilt 2010: 54). In 1919, US Secretary of War, Newton Baker, warned that the American bombing of civilians would constitute 'an abandonment of the time-honoured practice among civilised people of restricting bombardment to fortified places...' (1919). Baker's position was grounded in the conviction that the materially non-threatening had a basic right to be free of direct and deliberate targeting.

The League of Nations Disarmament Conference between 1932 and 1934 (LON 1932) featured a second attempt to prohibit, rather than merely regulate the practice of aerial bombing. These efforts were led by Britain, largely for military and economic reasons.[25] Importantly though, Britain's position also had a normative dimension, grounded in their opposition to the bombing of the defenceless. As one Internal Memorandum from the British Admiralty argued:

> Air bombing is very aggressive and in no way defensive...The Army and Navy do not want [bombers]. Only the Air Ministry wants to retain these weapons for use against towns, a method of warfare which is revolting and un-English. (1932, cited in Bialer 1980: 24)[26]

While efforts continued throughout the 1930s to implement a general ban on aerial bombing, no agreements were made, or sought thereafter.[27] The consequences of this were laid bare during the Second World War, a conflict that

[25] In addition to the financial strain and instability that would follow an aerial arms race, Britain feared a 'bolt from the blue'—a surprise attack by aircraft that would nullify its traditional naval protections (Biddle 2014: 37; Meilinger 2008: 79). This opposition was undermined by Britain's insistence that it retain the right to bomb indigenous peoples as part of its colonial 'policing duties'. In response to protest, Britain eventually agreed to abandon this exception (Biddle 2014: 37).

[26] During the League of Nations Disarmament Conference, the 'Hoover Plan' was also introduced by the United States, which went as far as to advocate not only a prohibition against aerial bombardment, but also an abolition of all bombing aircraft. This proposal was rejected. A major impediment to the plan was civil aviation, and the belief that its development would render restraints on military aircraft meaningless (Meilinger 2008: 80).

[27] British legal expert J. M. Spaight argued in 1938 that the laws relating to aerial bombing were 'in a state of baffling chaos and confusion which makes it almost impossible to say in any given situation what the rule really is' (1938: 25). Following the Second World War, further attempts were made to address this perceived shortfall in the laws of war regarding aerial bombing. Boothby, however, argues that it was not until the ratification of the 1977 API that formal international agreement was reached on the rules governing 'the engagement from the air of targets on land' (2014: 218–19).

'reduced to the vanishing point the protection of the civilian population from aerial bombardment' (Lauterpacht 1952: 529). While insufficient regulation had a role in this breakdown, ultimately it was the *indivisibility* of total war that guaranteed it. Anscombe summarized this logic thusly:

> The civilian population we were told, is really as much combatant as the fighting forces. The military strength of a nation includes its whole economic and social strength. Therefore the distinction between the people engaged in prosecuting the war and the population at large is unreal. There is no such thing as a non-participator; you cannot buy a postage stamp or any taxed article, or grow a potato or cook a meal, without contributing to the 'war effort'. War indeed is a 'ghastly evil', but once it has broken out no one can 'contract out' of it... [this] was a doctrine of 'collective responsibility'... (1981a: 63)[28]

Total war pressured and overturned each of the dominant accounts of Just War. Within these conditions, the enemy was fully collectivized—all individuals, regardless of status, were considered a constitutive element of the overall material threat. Even if taken individually, however, the causal contribution of non-combatants was judged to cross the necessary threshold for liability to harm. Lastly, total war overturned the implicit 'contract' of armed conflict, the consequence of which was a collapse of the traditional targeting distinction between protected non-combatants and liable combatants.

Opposition to this bombing was extensive. Jesuit priest and Just War thinker John C. Ford was one such critic. In his famous essay, 'The Morality of Obliteration Bombing', Ford criticized the targeting of civilians on both necessity and liability grounds:

> [I]t is true that the number of civilians who contribute immediately to the armed prosecution of the war has increased in modern times, but to say that all or nearly all do so is a grave distortion of the facts. (1944: 283)

Ryan opposed the targeting of civilians from the air for similar reasons, writing that even within conditions of total war, 'the co-operation, moral and physical, of the generality of men, women, and children is not so immediate in time, place, or character as to give them the same essential status as active combatants' (cited in Ford 1944: 278–9). The position of Ford and Ryan was not that civilians contributed nothing to the collective threat of their side,

[28] See also, Douhet (2009: 196).

they clearly did. Crucially though, this contribution was of a level *insufficient* for moral liability. It was morally indefensible, therefore, to target them. Implicit within this criticism was the rational presumption that combatants—all combatants—*did* pass this liability threshold.

The ruinous wars of the first half of the twentieth century illustrate the depth of the historical failure to prevent the aerial bombing of civilians. Yet as this section has outlined, opposition did exist to the practice, grounded in the belief that the materially non-threatening were not liable to be harmed. It has been the opposite presumption (that combatants *do* sufficiently contribute to the material threat of their side) that has historically immunized the bombing of ground-based combatants from the same degree of moral opprobrium.

This traditional division between targetable and protected came under strain in the 1990s. During both the First Gulf War and OAF, an ethical tension emerged as a consequence of the near-riskless status of US pilots. Crucially though, this same asymmetry also undermined the presumption of material *threat* among the combatants of the weaker side, to a degree that called into question the permissiveness of such violence—a challenge that had hitherto arisen only in relation to civilian targeting. The next section will examine this in greater detail, beginning with Operation Desert Storm (ODS) of the First Gulf War. Particular focus will be given to the Highway of Death episode.

The Antecedents of Radical Asymmetry

The First Gulf War: Highway of Death

ODS commenced on 17 January 1991, in response to the invasion and annexation of Kuwait by Iraqi forces. The UN-authorized, US-led operation spanned forty-three days, with the first thirty-nine prosecuted exclusively through air and naval bombing. The bombing component of ODS offered an early glimpse of a new mode of casualty-averse and technocentric Western war, in which airpower replaced, rather than merely augmented, the use of ground troops. What followed was an imposition by US forces of one of the most significant differentials of risk in the history of warfare. At the conclusion of ODS, Iraqi forces had suffered approximately 100,000 combat deaths, compared with an estimated 140 (and 366 casualties) on the side of the US-led coalition (Münkler 2003: 10).[29] To properly comprehend this disparity in

[29] During the conflict, the United States found surprising ways to maximize their aerial advantage. They soon discovered, for example, that Iraqi armoured vehicles were easily detectable on infrared

historical terms, during the Vietnam War there were 127,405 US casualties, translating into a casualty rate of 0.3 per cent of daily overall theatre strength. During the Gulf War, this rate had dropped to 0.0016 (cited in Riza 2013: 43). According to two sociologists, the death toll of the United States represented only 40 per cent of the number that might otherwise have died domestically during the same period of time (Garchik 1991: A8). The United States effectively saved American lives by going to war (Gray 1997: 37).

There were many, however, who questioned the ethics of this military imbalance, suggesting that US forces had divorced themselves from risk to such a degree as to problematize their status as warriors. French philosopher Jean Baudrillard advanced this argument in a collection of essays titled *The Gulf War Did Not Take Place* (1991). The title was in reference to what Baudrillard perceived as the conclusion of war as a 'confrontation of warriors' (1991: 86). What had emerged in its place was a techno-war of distance: 'clean war, white war, programmed war', but crucially, one 'more lethal than the war which sacrifices human lives' (Baudrillard 1991: 56). For some combatants, this absence of risk translated into an extra sensitivity regarding their own battlefield conduct. Glover writes that when British army helicopters were ordered to fire upon retreating Iraqi tanks:

> They would circle over the tanks as a warning to the crews, who could escape before the tanks were destroyed…'*that way we could feel at peace with ourselves*. We had total superiority, but we didn't use it'.
>
> (2001: 53, my emphasis)

The influence of the warrior ethos could also be seen in the eventual decision by US policymakers to terminate the conflict. According to then Deputy National Security Advisor Robert Gates, Colin Powell had argued that the use of American violence against Iraqi forces beyond a certain point would be 'un-American', and 'unchivalrous' (cited in Rumsfeld 2011: 413–14). Here we find an illustration of the same ethical tension that had long been a feature of civilian bombing. Crucially, though, as a result of the degree of asymmetry that characterized the First Gulf War, the scope of this challenge had expanded to include *inter-combatant* violence. Of even greater significance was the

displays between sunset and midnight due to the fact that their rate of heat dissipation was slower than the surrounding desert sand (Lambeth 2000: 124). This allowed American forces to conduct 'tank plinking' attacks at night, at a reduced risk to the pilot. For more detail on the technical aspect of this bombing, see Allen, Berry, and Polmar (1991: 72).

suggestion that this same asymmetry challenged not only the warrior ethos, but also the permissive moral standards of the Just War Tradition.

The risk imbalance of the First Gulf War was determined just as much by the weakness of the Iraqi military as it was by the strength of the United States. According to one US Army analyst, Iraqi troops were 'ill-trained and ill-prepared to execute this kind of defence' (Pardew 1991: 21). 'They were like civilians thrown into a military environment', argued another (cited in Mueller 1995: 101). 'On a combat scale of 1 to 10', claimed one US Marine, the Iraqi army 'was a 1' (cited in Mueller 1995: 106–7).

This weakness was brought into stark relief during an incident that was subsequently termed the Highway of Death. During this encounter, which took place a day before the formal conclusion of hostilities, the material threat of Iraqi forces was minimized to such a degree as to call into question the permissiveness of American violence. Iraqi military personnel retreating from Kuwait along Highway 80 were blocked by American and Canadian forces and subjected to ten hours of bombardment. In the process, Western forces destroyed 34 Iraqi tanks, 41 armoured personnel carriers, 43 artillery pieces, 319 anti-tank guns, 224 trucks, and killed 400 fighters, with no loss of Coalition life (Newton and May 2014: 73–4).[30]

Even when viewed alongside the significant military imbalance that preceded it, the asymmetric violence of the Highway of Death was distinct. *The New York Times* characterized the episode as a 'turkey shoot' (Sciolino 1998). US Squadron Leader Frank Sweigart described the fleeing Iraqis 'as basically just sitting ducks' (cited in Burrell 2011: 187). According to other media sources, the incident was 'arguably more slaughter than war' (Goldman and Matthews, cited in Mueller 1995: 115). The *Telegraph* termed it simply: 'a massacre' (cited in Pilger 1994: 152).[31]

Given these descriptions, it is perhaps unsurprising that the episode generated significant controversy. Of particular concern was the apparent absence of material threat posed by the targeted. The resulting rout, argued Hughes, 'troubled many humanitarians who otherwise supported the allied objectives' (cited in Cockburn 1991). This included Walzer, who suggested that 'standard just war arguments would probably come down against bombing the chaotic flight from Kuwait…' (1992: 14).

[30] A similar, but less publicized attack also took place on Highway 8.

[31] It should be noted that these claims were contested. John Mueller is among those who argue that American violence did not constitute a 'mass slaughter' (1995: 111).

As Chapter 4 outlined, within both conventionalist and revisionist accounts of Just War, the permissiveness of inter-combatant violence is contingent on the presumption that combatants *sufficiently* contribute to the collective material threat of their side. What made the Highway of Death episode morally problematic for so many was the extent to which the coherence of this presumption was undermined. The material threat of Iraqi combatants had been rendered so marginal as to blur the liability distinction between them and protected non-combatants (Lazar 2015: 117). '[It] would be hard', wrote May, 'to defend such killing on the grounds of the self-defence rights of the coalition forces' (2015b: 93). For Colin Powell, the asymmetry of violence made 'an obscenity of that aspect of Operation Desert Storm' (cited in Gaita 2003: 97). 'We don't want to be seen as killing for the sake of killing', Powell told President Bush (cited in Divine 2000: 133).[32] Comparisons can immediately be made to the opposition directed against military sniping, and particularly, the targeting of sentries during the American Civil War. In a notable contrast, however, the asymmetry of the Highway of Death went beyond the threat-status of specific targets, challenging the presumption of enemy threat across the full scope of the battlefield.[33]

Some went as far as to question whether the incident, on account of its asymmetry, could even be understood as *war*. In his own writing on the attack, Walzer observed that:

> One may object to killing in war, even in just war, whenever it gets too easy. A 'turkey shoot' is not a combat between combatants. When the world divides radically into those who bomb and those who are bombed, it becomes morally problematic, even if the bombing in this or that instance is justifiable. (1992: 14)

[32] This mentality was evident in some of the earlier rhetoric of US commanders. On 14 November 1990, former US Army general and head of coalition forces, H. Norman Schwarzkopf, stated: '[w]e need to destroy – not attack, not damage, not surround – I want you to destroy the Republican Guard...We're not gonna say we want to be as nice as we possibly can, and if they draw back across the border that's fine with us. That's bullshit! We are going to destroy the Republican Guard' (1992: 381–2). In a subsequent interview with *Newsweek*, Schwarzkopf did, however, wonder: 'how long...the world [would] stand by and watch the United States pound the living hell out of Iraq without saying, "wait a minute – enough is enough"' (cited in Newsweek 1991).

[33] Though a minority position, some suggested that the 'Highway of Death' was not merely morally troubling, but also a war crime, and specifically, a violation of the necessity standard of the laws of war. Kenneth Roth, Deputy Director of Human Rights Watch at the time, claimed that 'it wasn't a tactical retreat...It was a panicked, desperate flight and these were just lowly soldiers trying to get the hell out of there. It appears that this was a case of senseless slaughter' (cited in Waller and Barry 1992: 16). This was echoed by at least one American lieutenant general, who later described the attacks as controversial and disproportionate (Sanchez and Phillips 2008: 80). Chediac argued that '[t]he massacre of withdrawing Iraqi soldiers violate[d] the Geneva Conventions of 1949, Common Article III, which outlaws the killing of soldiers who are out of combat...' (1991).

As this further highlights, the risk imbalance that characterized the Highway of Death also strained the coherence of the contractarian position, which hinges on a mutuality of risk between contesting adversaries. The conflict had instead become a one-sided dispensation of violence; 'a giant hunt', as simple as 'shooting fish in a barrel' (Aksoy and Robins 1991: 333).

The First Gulf War marks an important turning point in the asymmetry-challenge of manned aerial bombing. The military advantage of the stronger party was for the first time deemed so overwhelming as to call into question the moral justifications for the permissiveness of inter-combatant violence. In the years that followed, these questions grew only more pressing with the arrival of 'zero combat casualty' warfare.

Operation Allied Force: Zero Combat Casualty Warfare

OAF refers to the NATO bombing of the Serbian-dominated Federal Republic of Yugoslavia (FRY). The conflict began on 24 March 1999 and was prosecuted for seventy-eight days, with the intention of halting ethnic cleansing in Kosovo and compelling a withdrawal of Serb forces from the disputed province. According to then President Bill Clinton, this was to be achieved through the concentrated targeting and destruction of Serbia's military capacity (1999). While ostensibly a NATO campaign, formally supported by all nineteen member states, the mission was dominated by the United States.[34] One of the most notable features of the campaign was the early decision by Clinton to rule out the use of ground forces (Record 2000: 4). Prior to hostilities, Gen. Hugh Shelton stated that as many as 200,000 ground troops would be necessary to overcome Serbian resistance (cited in Meilinger 2003: 110). Instead, and somewhat paradoxically, NATO ground troops were to be utilized only after air strikes had secured a 'permissive environment'. This was understood as the complete termination of hostilities, the withdrawal of all Serb forces, and a signed agreement from President Milosevic allowing Kosovar Albanian refugees to return to Kosovo (BBC News 1999).

Two key reasons explain NATO's rejection of ground forces and exclusive reliance on air power as the instrument of armed rescue in Kosovo. The first

[34] Between 24 March and 10 June, the United States conducted over 60% of all air and missile sorties, 53% of strike-attack sorties, and 71% of overall support sorties (Wheeler 2000: 266). The United States also flew over 90% of electronic warfare and advanced intelligence and reconnaissance missions; and launched over 95% of cruise missiles and 80% of strike-attack munitions and precision guided air weapons (Cordesman 2001: 64; Wheeler 2000: 266). For additional figures on the bombing, see Balabanova (2007: 45).

stems from the conviction of Western leaders that a sufficiently serious demonstration of airpower would compel Milosevic to capitulate after only a few days of bombing (Wheeler 2000: 268).[35] The second and most important reason was the belief, particularly among US policymakers, that soldiers returning in body bags would rapidly erode domestic support for OAF (Wheeler 2000: 268).[36] With ground troops ruled out, OAF holds the distinction of being the first war to be waged exclusively through the use of air strikes. These strikes were further notable in that pilots were primarily restricted to an attacking altitude of 15,000 feet for the duration of the conflict.[37] This creation of high-altitude 'sanctuaries' positioned US forces out of range of most hand-held surface-to-air missiles and anti-aircraft artillery (Robinson 1999: 13; Rogers 2000; Truver 2000: 35). The combination of the total absence of US ground forces and high-altitude bombing generated a disparity of risk without historical precedent. Excluding accidents and friendly fire, the 78-day air campaign over Kosovo and Serbia resulted in zero American combat casualties.[38]

As with the First Gulf War, the intensity of this military imbalance drew criticism from some, on account of its alleged incompatibility with the warrior ethos. Norman Mailer suggested that the conflict crossed an important ethical threshold. According to Mailer, more traditional ground war, though 'always cruel beyond human comprehension', nevertheless provided an opportunity for 'heroism and sacrifice'. The bombing of OAF, in contrast, governed as it was by 'the notion that our own blood is not to be shed... [was] obscene' (Mailer 1999: A25). Echoing this position, military historian Gwynne Dyer criticized the exclusion of ground troops and utilization of high-altitude bombing as 'morally defective... [a] coward's strategy' (1999: B3). For Woollacott, that the conflict had been 'bloodless' for NATO forces meant that victory had not been 'earned' (1999). Former French general and Commander of UN Forces in Bosnia (1992–3), Philippe Morillon, questioned the commitment of the United States to zero-casualty warfare, asking: 'who

[35] This belief was influenced by the legacy of Bosnia, where Clinton, having ruled out ground forces, took away the lesson that Milosevic had been 'bombed' into the Dayton agreement. This mythologizing of the 1995 bombing overlooked two important factors—first, that it followed a period of significant Serb military reversals on the ground; and second, that the 1995 bombing was not directed against Serbia proper, and thus did not arouse the same nationalistic response as the bombing of 1999 (Roberts 1999: 110).

[36] Former British Prime Minister Tony Blair was alone among Western leaders in his support for a ground operation in Kosovo (Evans 1999).

[37] Despite claims to the contrary, there were occasions in which pilots operated at lower altitudes (Dunlap 2014: 111).

[38] As clarified in the introductory chapter of this book, two American servicemen were killed during a night-time training exercise (Graham and Drozdiak 1999: A1).

are these soldiers, who are ready to kill but not ready to die?' (cited in Beevor 1999). Former British officer Paul Robinson went further, suggesting that the near-risklessness of OAF degraded the ethical standing of NATO forces, transforming them from 'warriors into mere killers' (1999: 673).[39]

Alongside this tension with the warrior ethos stood a potentially deeper challenge, grounded in the *moral* status of US airstrikes. For many observers, American violence had advanced beyond any reasonable claim of self-defence. 'The expectations that govern the morality of war', suggested Ignatieff, 'grounded in a basic equality of moral risk' were fundamentally jeopardized by the one-directional nature of American violence (2000: 161). This was echoed by Kahn, who argued that the United States had nullified reciprocal risk to such a degree, it could no longer draw upon traditional moral justifications for lethal force (2002: 2).

Recall, however, the critical distinction between the role of reciprocal risk within the warrior ethos and its function within the Just War Tradition. In the former it contains an inherent value, given the centrality of personal risk in the ethical legitimacy of violence. Within the Just War Tradition, in contrast, reciprocal risk is necessary only to the extent that it marks both sides as mutual material threats, and on that basis, mutually liable for targeting. Crucially, though reciprocal risk had been almost fully undermined between the United States and Serb forces, the latter remained an unambiguous threat. As former President of the Czech Republic Vaclav Havel argued:

> This is probably the first war ever fought that is not being fought in the name of interests, but in the name of certain principles and values . . . if it is possible to say about a war that it is ethical, or that it is fought for ethical reasons, it is true of this war . . . Milosovic is not threatening . . . the territorial integrity . . . of any NATO member. Nevertheless, the Alliance is fighting . . . for the fate of other human beings . . . (1999).

Whatever the sincerity of this motivation, or the extent to which these stated goals were realized, there is little doubt that Serb forces presented a very real and ongoing threat to the Albanian Kosovars on the ground. Because of this, and in spite of the asymmetry between American pilots and Serb forces, the moral liability of the latter to lethal targeting endured.

The concluding section of this chapter will address the resolution of the asymmetry-challenge of manned aerial bombing. It argues that the moral

[39] For similar perspectives, see Evans (1999); and Tilford (1999: 37).

significance of both ODS and OAF was ultimately constrained by the failure of the United States to sufficiently undermine the structural reciprocity of war.

The Limits and Resolution of the Challenge of Manned Aerial Bombing

Today, the manned aerial bombing of ground-based combatants, if conducted in accordance with the explicit *in bello* rules of war, is accepted by most as morally justified, regardless of the risk-disparity between pilot and target. What then explains the failure of the opposition listed above to find lasting historical purchase? This chapter argues that the principal explanation for this impermanence is that while both ODS and OAF featured unprecedented military imbalance, neither sufficiently undermined war as a site of reciprocal structural risk. The coherence of the moral frameworks that justified the permissiveness of American violence, though strained, ultimately remained intact. Consequently, the resolution of both challenges can be explained primarily in relation to the warrior ethos, and specifically, the growing importance of combat responsibility within the framework itself.

The Individual and Structural Risk of Manned Bombing

It is first important to note that despite the historically enduring sentiment that aerial bombing imposed a problematic degree of reciprocal risk, bombing, like sniping, has always been a highly dangerous pursuit. Those who looked upon the bombing of the two world wars as cowardly, Wragg argues, 'had little conception of the risk involved and of the casualty rate amongst bomber aircrew' (1986: 14). The Allied pilots that bombed German cities faced constant threats, including icing, mechanical failure, anti-aircraft batteries, and Luftwaffe fighters. As a result, British bomber crews suffered the highest attrition rates of any branch of the British armed forces during the Second World War (Lee 2012: 9–10). In some circumstances, pilots volunteered to enhance these risks even further. When forced to bomb their own cities during the Second World War, some French pilots flew at riskier, lower altitudes, in order to minimize French civilian casualties (Walzer 2006a: 157).

By the time the First Gulf War was fought, the physical risk experienced by American pilots had been dramatically reduced. It had not, however, been eliminated. The first American combat casualty of the war was a Navy FA-18 fighter-bomber pilot, killed the first night of ODS by Iraqi forces, most likely via a surface-to-air missile (CIA 2001). Of even greater significance for the moral status of American violence was the fact that at no point did the *structural* threat posed by Iraqi forces fully conclude. Even during the Highway of Death incident, the presumption that Iraqi combatants on the ground maintained the capability to threaten attacking Coalition forces was a defensible one. The permissiveness of American violence, therefore, remained justifiable according to the moral standards of both conventionalist and revisionist accounts.[40] The ongoing risk to American forces, though minimized, was also sufficient to signal the maintenance of the implicit adversarial bargain at the heart of the contractarian right to kill.

As was noted earlier, in the case of OAF, the material threat posed by Serb forces to third parties was sufficient to render them morally liable to lethal targeting on the battlefield. Crucially though, even if we ignore this reality and focus exclusively on the military imbalance between the American and Serbian military, structural reciprocity endured. Though NATO forces experienced zero combat casualties for the duration of the conflict, this outcome was not guaranteed. Despite the extraordinary measures implemented to avoid casualties during the campaign, policymakers and commanders did expect some loss of American life (Leaf, cited in Lyon 2001: 61; Short 1999). OAF was deemed riskless only after the fact. This is logical, as the physical presence of human fighters, no matter the altitude at which they operate, invariably translates into at least some measure of risk. Though tested, the moral standards that justify the permissiveness of inter-combatant violence retained their coherence during this conflict.[41]

Both the First Gulf War and OAF offered an early glimpse of the potential of risk-disparity to challenge not only the warrior ethos, but also the very coherence of the moral justifications for killing in war. The endurance of structural reciprocity within both conflicts, however, was ultimately sufficient to ensure the moral permissibility of American violence. Just as with sniping,

[40] This was equally true in legal terms. Dinstein rejects as a 'serious misconception' the claim that retreating Iraqi forces were *hors de combat* (2002: 153). According to Dinstein, the *lex lata* rules of war would only have prevented attack if Iraqi forces had formally surrendered (2002: 153).

[41] So too did the laws of war. According to the ICTY: 'there...[was]...nothing inherently unlawful about flying above the height which can be reached by enemy air defences' (Ponte 2000: Par 56). An Amnesty International Report did contest the legality of NATO strikes, writing: 'the requirement that NATO aircraft fly 15,000 feet, made full adherence to international humanitarian law virtually impossible' (2000: 25). This claim, however, was in reference to the alleged inability of US pilots, by virtue of their altitude, to properly discriminate between targets.

therefore, the resolution of the asymmetry-challenge of manned aerial bombing was ultimately determined in reference to the warrior ethos alone.

Combat Responsibility and the Aircrew Ethos

Chapter 3 presented the warrior ethos as an adaptive ethical framework, one that has evolved to reflect the changing material and ideational character of war. This adaptive potential stands as a key factor in the ethical normalization of manned aerial bombing:

> [The ethos] is not formed in a conceptual vacuum but draws upon familiar, existing discourses, incorporating those aspects that are relevant, or can be modified to be relevant, in the present while ignoring those concepts from the past that are not. (Lee 2012: 3)

As it has evolved, the aircrew ethos has de-emphasized personal, physical risk, and established combat responsibility as a principal component in the ethical legitimacy of bombing. Within the American context, this shift in ethical focus was particularly pronounced in the aftermath of the Vietnam War. In relation to the practice of indiscriminate bombing, former Secretary of Defence Robert McNamara noted at the time:

> There may be a limit beyond which many Americans and much of the world will not permit the United States to go. The picture of the world's greatest superpower killing or seriously injuring 1000 non-combatants a week, while trying to pound a tiny backward nation into submission on an issue whose merits are hotly debated, is not a pretty one. (cited in Gibson 1986: 368)

These fears proved well founded. The 'systematic exposure of Vietnamese civilians to the violence of American warmaking' oxygenated the war's domestic opposition and severely diminished the moral credibility of the United States (Walzer 2004c: 7). In the years following, the principle of non-combatant immunity was systematically internalized into the military doctrine and training procedures of the United States, as well as the development of new weapon systems (Kaempf 2009: 666). This commitment to restraint was evident throughout both the First Gulf War (Thomas 2001: 88) and OAF. In relation to the latter conflict, US Gen. Chuck Wald claimed that the rules of engagement for US pilots were 'as strict as I have seen in twenty-seven years'

(E. Schmitt 1999).[42] The degree to which US pilots adhered to the rules and standards of war helped mitigate, if not completely offset, the ethical tension that emerged as a consequence of their asymmetric advantage.

The willingness of pilots to endure the risk of injury and death remains an important component of the aircrew ethos. This was confirmed by Barrett in a 1996 study of hegemonic masculinity within the US Navy. Barrett argued that the construction of masculine identities among naval aviators 'drew upon themes of autonomy and risk-taking' (1996: 129). For such risk to be worthy of veneration, however, it must be coupled with combat responsibility. Within its *Ethos, Core Values and Standards*, the Royal Air Force states that 'courage, *both physical and moral*, forms the bedrock upon which bravery, fighting spirit and success depend' (2008: 6, my emphasis). The ethos of the Australian Air Force places equal value on combatant restraint. In 2015, it was reported that Australian pilots engaging targets in Northern Iraq aborted one in seven bombing runs due to an unacceptable risk to civilians (Nicholson 2015). According to US Air Force Chief of Staff Gen. Mosely, the 'warrior ethos exhibits a hardiness of spirit, and moral and physical courage' (cited in Gettle 2007: 5). Mosely further observes that:

> Precision air strikes against the enemy save American and coalition lives. Taking out a target with a precision-guided missile or bomb, versus sending troops in to take out the same target, saves valuable ground forces... Our Airmen bring more to the fight today than ever in Air Force history, but we do it in a way that puts our people in less danger. (cited in Gettle 2007: 5)

When the ethical status of low-risk aerial bombing is called into question today, criticism is rarely grounded in the notion that such violence is insufficiently courageous. The concern is instead that Western supremacy in the air is too often obtained at the expense of a necessary commitment to non-combatant safety. The recent decision by Western Coalition forces to pursue an air-exclusive campaign against ISIS has provoked widespread criticism from those troubled at the possibility that such tactics imperil the safety of civilians on the ground (Khan and Gopal 2017). It is restraint and rule-compliance, not physical risk, that determines the legitimacy of contemporary manned bombing.

[42] For more detail on the discriminate nature of OAF, see Farrell (2005: 159); Kaempf (2009: 666); Shaw (2005: 21); and Wheeler (2004: 198). The moral and legal status of this bombing will again be assessed in the concluding chapter.

The aerial fighting of the two world wars was recognized as a chivalric expression of warrior virtue. Central to this ethical status was the pilot's relationship with an extreme and personalized form of danger. Within the contemporary aircrew ethos, physical risk, though maintained, has been de-emphasized. Ethical status is instead crafted around a narrative of 'protection'—the safeguarding of both enemy non-combatants and ground-based allies. The ethical normalization of aerial bombing has not required the abandonment of physical risk as a feature of ethical killing. Increasingly though, such risk is preconditioned on the exercise of a morally and legally restrained mode of violence. This, coupled with the endurance of structural reciprocity, explains the resolution of the asymmetry-challenge of manned aerial bombing.

Conclusion

Historically, few modes of violence have distorted the reciprocal risk of battle to a greater degree than manned aerial bombing. Just as with sniping, aerial bombing could be conducted at a distance that severely reduced the capability of those targeted to physically retaliate. As this chapter has illustrated, however, aerial bombing is distinct from sniping in that the asymmetry imposed by the former was regarded for most of history as problematic only when directed against non-combatants. This remained the case until the 1990s, when the locus of this challenge expanded for the first time to include the bombing of combatants. Within the First Gulf War—and in a more qualified sense, OAF—aerial bombing generated an asymmetry of risk significant enough to call into question the permissiveness of American violence. Crucially though, the endurance of a necessary degree of structural risk meant that the coherence of the moral justifications for killing in war, though tested, ultimately endured.

Problematically, it is precisely this structural reciprocity that is increasingly challenged within the contemporary context of radically asymmetric violence. Chapter 7 argues that UAV-exclusive violence has the potential to mitigate inter-belligerent risk to a degree that imperils the coherence of the existing moral frameworks that justify lethal force. This marks a qualitative disjuncture from the historical challenges of asymmetry examined so far.

7
UAV-Exclusive Violence

Introduction

Throughout the history of war, modes of violence have emerged that have mitigated reciprocal physical risk between opposing combatants to a degree sufficient to challenge the expectations of the warrior ethos. Chapters 5 and 6 examined two of these historical episodes, in relation to military sniping and manned aerial bombing respectively. Within both cases, an asymmetry of risk was imposed upon the battlefield that called into question the ethical legitimacy of such violence.

At the same time, there exists a more fundamental threshold of military imbalance, located at the *structural*, inter-belligerent level of war. If crossed, such imbalance has the potential to undermine the very coherence of the moral justifications for lethal force. This chapter argues that UAV-exclusive violence has crossed this threshold, and in doing so, strained the fundamental assumptions upon which the permissive moral standards of inter-combatant violence rest.[1] 'UAV-exclusive' refers to the ongoing use, most notably by the United States, of UAVs in the near-total absence of military alternatives, including ground-based forces. In short, UAVs *as* war, rather than *in* war. The imposition of this radically asymmetric mode of violence, as seen at certain times in theatres such as Pakistan, Yemen, and Somalia, threatens to erode the capacity of the United States to interpret and apply standard judgements of Just War theory against those it targets.

This chapter will first provide a historical overview of the UAV, commencing with the earliest versions of the technology and concluding with an assessment of its current asymmetric potential. It then explores the ethical tension exposed by this imbalance of risk. It argues that, just as with military sniping and manned aerial bombing, the asymmetry-challenge of UAVs derives partly from the insufficient exposure of its operators to physical risk.

[1] The US DoD defines UAVs as follows: 'A powered, aerial vehicle that does not carry a human operator, uses aerodynamic forces to provide vehicle lift, can fly autonomously or be piloted remotely, can be expendable or recoverable, and can carry a lethal or nonlethal payload' (Bone and Bolkcom 2003: Summary). This chapter focuses exclusively on the lethal variants of this technology.

Asymmetric Killing: Risk Avoidance, Just War, and the Warrior Ethos. Neil C. Renic, Oxford University Press (2020).

DOI: 10.1093/oso/9780198851462.001.0001

In further parallel with these historical episodes, however, this tension with the warrior ethos has a high likelihood of resolution. Of greater concern are the *moral* dimensions of this challenge. The chapter argues that the coherence of each of the dominant positions of Just War is problematized within conditions of radically asymmetric UAV violence, particularly when directed against low-level enemies. The chapter concludes with an assessment of the broader implications of this moral challenge. It argues that in addition to incentivizing those targeted to devote greater resources to criminal methods of retaliation, UAV-exclusive violence marks a fundamental shift in the nature of hostilities, from an adversarial contestation to something more closely approximating judicial sanction. One outcome of this shift has been the dehumanization of those targeted by the United States.

Historical Overview

Aviation history was made in July 2013 when the US Navy's X-47B became the first UAV to successfully complete an autonomous landing on an aircraft carrier. Capable of achieving speeds of 500 miles per hour at an altitude of 40,000 feet, the fully autonomous X-47B was designed to execute pre-programmed attack missions without human intervention (Alexander 2013a). The landing, announced Rear Admiral Ted Branch, represented a 'marker…between naval aviation as we've known it and the future of naval aviation…' (cited in Alexander 2013a).[2] The following section examines the conceptual and material factors that have made this future possible. Particular focus is given to the gradual increase in the asymmetric potential of this mode of violence.

Early Concepts and Designs

One of the earliest proponents of the UAV was Serbian-American inventor Nikola Tesla, who in an 1898 article detailed his vision for a remotely controlled aircraft. Titled the 'teleautomaton', this highly manoeuvrable weapon was intended to kill with a precision that exceeded the military alternatives of the period (1898). Less than twenty years later, a version of this concept was

[2] In 2015, the US Navy announced it would retire the X-47B, and field the Unmanned Carrier-Launched Airborne Surveillance and Strike System (UCLASS) by 2022 (LaGrone 2015).

put into practice with the 'Curtiss-Sperry Flying Bomb'. Supported by the US Navy Bureau of Ordnance, this precursor to both the cruise missile and UAV was intended for use against German U-boat bases during the First World War. Tests in 1917, however, proved only partially successful and the weapon was never deployed in battle (Pearson 1969: 70–2). During this same period, the US Army tested the 'Kettering Bug', an unmanned vehicle capable of carrying a 250-pound warhead. Just as with the 'Curtis-Sperry Flying Bomb', the First World War had concluded before it was utilized in combat (Newcome 2004: 29–30; Spinetta 2010).

A more successful antecedent of the UAV was the German V-weapon programme of the Second World War. Known as the 'Vengeance Weapons', this included the V-1 and V-2 rocket, as well as the V-3 cannon (Keane and Carr 2013: 564). The V-1, a winged, pilotless fuel-propelled flying bomb, was first launched against London on 12 June 1944 (Chant 1996: 44). Though the military impact of these weapons was comparatively modest—the V-1 and V-2 rockets were collectively responsible for 2,500 British deaths—they nevertheless served to underscore the potential of unmanned warfare.[3] Gen. Hap Arnold, then Chief of the US Army and Air Force, gave voice to this optimism in 1945:

> We have just won a war with a lot of heroes flying around in planes. The next war may be fought by airplanes with no men in them at all…tomorrow's aviation…will be different from anything the world has ever seen.
>
> (cited in Spinetta 2010)[4]

The pursuit of robotics in war has long been imbued with a sense of *deus ex machina*. Conceived as a 'silver-bullet', unmanned technology has the capacity to divorce practitioners from the reciprocal vulnerability of the battlefield. This potential benefit was the subject of greater scrutiny in a 1964 study by engineer John W. Clark, titled 'Remote Control in Hostile Environments' (1964: 300–3). 'It is unnecessary', wrote Clark, 'to require a man to expose himself to physical danger in order to earn a living' (1964: 300). The solution, he argued, lay in 'telechiric machines', a somewhat cumbersome term to

[3] This low figure, at least in comparison to manned aerial bombing, is partly owed to the late emergence of the superior V-2 rocket, as well as a number of limitations with the technology itself (Chant 1996: 44).

[4] A year later, *The New York Times* published a story on the potential tactical benefits of armed pilotless planes (Baldwin 1946: 96). In contrast to this emphasis on the lethal application of this technology, actual advancements at this time steered more towards surveillance, as evidenced by the emergence of the radar-tracked SD-1 'recon-drone' (Palmer 2008: 10).

describe the 'technology of manipulation at a distance' (Clark 1964: 300). Clark's principal focus, however, was the potential application of these machines in hostile non-combat environments, such as outer space, nuclear zones, and the deep ocean (1964: 300). Although Clark conceded that these devices could be of use in war, his imagination went no further than machine-executed bomb disposal (1964: 303). At least one reader took issue with this seeming oversight, asking 'should not... [telechirists']... first efforts towards human safety be aimed at mankind's most hazardous employment – the industry of war?' ('Last Word on Telechirics' 1964: 405). The reader went on to describe this potential contribution with a prescience that is worth quoting at length:

> All conventional wars might eventually be conducted telechirically, armies of military robots battling it out by remote control, victory and defeat being calculated and apportioned by neutral computers, while humans sit safely at home watching on TV, the lubricating oil staining the sand in sensible simile of their own blood. Far-flung imperial conquests which were ours because we had the Maxim gun and they had the knobkerry will be recalled by new bloodless triumphs coming our way because we have the telechiric yeomanry and they, poor fuzzy-wuzzies, have only napalm and nerve-gas.
>
> ('Last Word on Telechirics' 1964: 405)[5]

The true battlefield potential of military robotics, however, was still some distance from being realized. Take, for example, the QH-50 Drone Antisubmarine Helicopter (DASH), which was equipped by the United States to a number of its destroyer warships in the 1960s. The experiment ultimately proved short-lived, with the arrival of destroyers large enough to launch manned helicopters. Despite this setback, the DASH stands as history's first hunter-killer UAV (Newcome 2004: 88).[6]

During the 1970s, interest in UAVs shifted from the United States to Israel, largely as a result of the range of tactical challenges faced by the latter. During the 1973 Yom Kippur War, the Israeli Defence Forces (IDF) had considerable success using unarmed UAVs to mislead Egyptian defences. Functioning as

[5] For an equally vivid account of this asymmetric potential, see an article published in 1973 by a group of young antiwar military scientists, which predicted the use of 'remotely piloted vehicles' in the suppression of 'Third World guerilla war'. 'Remote War', they argued, 'will defeat the human body. One side loses people; the other side loses toys' (Science for the People 1973: 39–40).

[6] For detail on other designs during this period, including the 'Ryan Firebee', D-21, and Compass Cope, see Yenne (2004: 35. 105).

an almost sacrificial wave, the UAVs allowed Israel's manned strikes to target the Egyptians as they were reloading (Chamayou 2015: 27; Keane and Carr 2013: 568).[7]

Israel also made an important technological contribution to the development of unmanned warfare, with its 1986 release of the RQ-2 'Pioneer', a UAV capable of reconnaissance and surveillance, as well as gunnery spotting (Newcome 2004: 99; Yenne 2004: 35). The Pioneer was operationalized by the United States during the 1991 First Gulf War, flying over 300 missions (Keane and Carr 2013: 569).[8] Arguably the most famous of these took place during the final week of the conflict. Following the successful use of the Pioneer by the USS *Missouri* to target its 16-inch gunfire against Iraqi defences, the USS *Wisconsin* sent its own Pioneer over the same defences at low altitude. Recognizing they were about to be targeted, five Iraqi soldiers waved white flags at the UAV's tiny television camera; 'it was the first time in history that men surrendered to a robot' (Coyne 1992, cited in Frontline 1996: 1). Antisubmarine warfare expert Henry Yuen wrote an internal paper shortly after the conflict, arguing the following:

> [O]ne of the foremost objectives in the development of new weaponry should be the reduction or total elimination of human risk. Put simply, weapons or equipment in harm's way should, to the extent possible, be unmanned. (cited in Toffler and Toffler 1993: 141)

More than any other military technology, the UAV has enabled the realization of this vision of riskless war. Today's UAVs, when utilized to the exclusion of on-the-ground human forces, impose conditions of radical asymmetry—lethal force actualized by one side in the near-complete absence of physical vulnerability, across the full scope of the battle space.

The Radically Asymmetric Potential of UAV Warfare

From its emergence in 1994, through until its formal retirement in 2018, the Predator stood at the forefront of unmanned warfare, particularly in the

[7] This tactic was replicated with similar success in 1982 against Syrian missile emplacements in the Bekaa Valley (Carus 1982).

[8] It was during this same period that the term UAV came into general use. From the Vietnam War up until this point, the common term had been 'remotely piloted vehicle' (RPV) (Newcome 2004: 1).

context of America's ongoing War on Terror.[9] Prior to the 11 September attacks, counterterrorism chief Richard A. Clark, along with then head of the CIA's Counter Terrorism Centre, Cofer Black, had speculated on whether to devise and employ armed UAVs in the hunt for Osama Bin Laden (Williams 2010: 872).[10] On 17 September 2001, President Bush authorized the CIA to target suspected terrorists globally. In response, Bush was presented with the 'Worldwide Attack Matrix', a plan for operations against terrorists in eighty states (Suskind 2007: 20–1). This included 'clandestine missions expressly aimed at killing specified individuals' (Gellman 2001: A01). Former CIA chief lawyer John Rizzo described it as 'the most comprehensive, most ambitious, most aggressive, and most risky Finding or MON [memorandum of notification] I was ever involved in' (2014: 173–4). On 7 October 2001, the United States launched Operation Enduring Freedom, and put its newly armed Predator UAVs to use hunting human targets in Afghanistan. The first of these lethal strikes is believed to have taken place in November of that same year.[11]

The actual use of UAVs by the United States is divided between two very different types of battlefield. The first are active 'hot' battlefields, such as Afghanistan, in which strikes have been conducted by the US military, in conjunction with ground forces and other kinds of air power. UAVs have also been deployed in the absence of a significant American ground presence, in regions such as Pakistan, Yemen, and Somalia. Operational command of this parallel campaign was initially split between two separate branches—Joint Special Operations Command (JSOC) and the CIA.[12] The first of these 'cold' UAV strikes took place in Yemen, in November 2002, against suspected al-Qaeda leader, Qaed Salim Sinan al-Harithi (Johnston and Sanger 2002).

[9] On 9 March 2018, the USAF formally retired the Predator and transitioned to the larger 'Reaper' UAV. For more information on the initial limitations of the Predator, see Cockburn (2015: 55). For detail on the design aspects of the Reaper, see Llenza (2011: 48).

[10] On 16 February 2001, the Predator was test fitted with Hellfire AGM-114c's, which it then successfully fired into a stationary tank, scoring three direct hits (Martin and Sasser 2010: 20).

[11] There are fewer details, however, regarding the intended target. Some media reports named Mohammed Atef, deputy to Bin Laden at the time (Fox News 2001; Yenne 2004: 90). Other sources listed Taliban Supreme Commander, Mullah Omar (Whittle 2014: 1), Defence Minister, Mullah Akhund (Woods 2011), or even Bin Laden himself (Sifton 2012).

[12] In response to growing concerns over what many perceived as the CIA's 'evolution from spy service to paramilitary force' (Miller 2015), the Obama administration announced in 2013 that it would shift much of its CIA-led UAV operations to JSOC control (Kaag and Kreps 2014: 34–5). This decision was subsequently reversed by President Trump (Evans 2017). For more detail on the distinction between these two programmes, see Blank (2015: 232); Boyle (2015: 118); Chesney (2016); and Williams (2011: 7).

Al-Harithi and six other suspected terrorists died in the attack.[13] A *Time* article published days after the strike described it as an 'assassination', but speculated that such violence was 'unlikely to become a norm' (Karon 2002).[14] To the contrary, these controversial strikes became an essential component of the American response to global terrorism. In a 2009 speech, then CIA Director Leon Panetta described UAVs as 'the only game in town in terms of confronting or trying to disrupt the al-Qaeda leadership' (2009). As we approach seventeen years since its initial use in Yemen, the application of UAV-exclusive violence by American forces has expanded to include Pakistan (Rohde and Khan 2004), Somalia (DeYoung 2011), the Philippines (Ahmed and Martin 2012), Libya (Turse, Moltke, and Speri 2018), and Niger (Reuters 2018).[15]

Though much of the contemporary debate over the use of UAVs has centred on the United States, this technology is becoming increasingly ubiquitous. Israel has deployed armed UAVs in both Gaza and the West Bank, including during the 2008–9 Operation Cast Lead (Mayer 2015: 767). The United Kingdom has also heavily increased its investment in UAV technology, while at the same time collaborating extensively with the United States (BBC News 2015; Woods 2015: 88–92). As of 2015, nearly eighty states were pursuing UAVs of some type, with approximately two dozen working towards long-range armed models (Mayer 2015: 769–70).[16] Part of the motivation for this growing investment is cost related. Predators are priced at US$4.5 million—thirty times less than a fighter jet (Sauer and Schörnig 2012: 370; Singer 2009: 34). Coupled with comparatively lower maintenance and training costs, UAVs represent an optimal military investment.

[13] Interestingly, the only survivor of the strike, Abdul Rauf Nassib, was apprehended in 2004, but acquitted by a Yemeni court over his alleged role in the USS *Cole* bombing. This information was revealed in a leaked US embassy cable (Wikileaks 2006).

[14] Only two months prior to the 11 September attacks, the United States had formally criticized Israel for its targeted killing programme. On 5 July 2001, then US ambassador to Israel, Martin Indyk, stated on Israeli television: '[t]he United States government is very clearly on the record as against targeted assassinations. They are extrajudicial killings, and we do not support that' (cited in O'Connell 2012).

[15] The extent to which US military action in these regions has been 'UAV-exclusive' has varied over time. The 2011 United Nations-sanctioned, NATO-led intervention against Libya, for example, clearly did not meet the criteria. Since 2011, however, the US has conducted approximately 550 drone strikes against Libyan targets.

[16] For more information on this expansion, see Fuhrmann and Horowitz (2017). Of potentially greater concern, the use of armed UAVs now extends to non-state actors. In October 2016, a UAV was rigged with explosives and launched by ISIS forces, killing two Peshmerga fighters and severely wounding two French soldiers in Iraq (The Guardian 2016). For more detail, see Larter (2017); and Ward (2017).

Of even greater significance is the potential of UAVs to impose conditions of radical asymmetry upon the battlefield. Western media has long framed the UAV and other unmanned weapons as 'life-saving technology' (Bray 2007; Chun 2006; Fleischauer 2006; Krasner 2007). According to a 2016 survey conducted by the Center for a New American Security, the possibility of alleviating air crew risk generated the strongest preference for unmanned warfare among the American public (2016). This preference also extends to the practitioners of American war. 'Air power', wrote Cohen, 'is an unusually seductive form of military strength, in part because, like modern courtship, it appears to offer gratification without commitment' (1994: 109). In the context of UAVs, this commitment, at least in terms of physical risk, is truly nullified. Put simply, unmanned warfare 'project[s] power without projecting vulnerability' (Deptula, cited in CNN 2009).

Military imbalance is of course nothing new. Throughout history, a range of technologies and practices have emerged that have allowed combatants to kill, while at the same time, limiting their exposure to the physical risk of battle. Yet in each of these cases, including military sniping and manned aerial bombing, 'a vestigial vulnerability has always remained: snipers can be shot or captured, planes can be shot down, and destroyers can be sunk' (Gusterson 2014: 196). Even if we consider one of the most extreme contemporary examples of asymmetry—the firing of cruise missiles from US Navy ships in the Mediterranean against targets in Libya, Iraq, and Sudan—there is a proximity between combatant and target that makes possible, however unlikely, a retaliatory response.[17] UAVs, at least in some cases, kill their target while their operator is as geographically removed from the encounter as it is possible to be. The extent of this distancing has advanced those empowered closer than ever before to the ultimate prize in war—riskless killing:

> By prolonging and radicalising pre-existing tendencies, the armed drone goes to the very limit: for whoever uses such a weapon, it becomes *a priori* impossible to die as one kills. Warfare, from being possibly asymmetrical, becomes absolutely unilateral... (Chamayou 2015: 12–13)

It should be noted that UAVs, particularly slower-moving and non-stealthy models, remain highly susceptible to the ground defences of most major

[17] For a legal defence of these strikes, see Delahunty and Yoo (2002: 507–12).

states.[18] While this has no bearing on the physical risk assumed by American operators, it does limit their potential application. At present, radically asymmetric violence is achievable only in a narrow context, when UAVs are utilized to the exclusion of significant ground forces, against non-state enemies. Within these specific conditions, however, the imbalance is profound. In many cases, those targeted are not only rendered non-threatening at the precise moment of their incapacitation or death, but are stripped—on aggregate—of their very *capability* to pose a credible physical threat to the military forces of the opposing side. This represents a transformative shift in the nature of war.

Just as with military sniping and manned aerial bombing, this asymmetry has a distinct psychological component. 'The drones were terrifying', wrote journalist David Rohde—who was kidnapped by the Taliban in 2008 and held for seven months; 'the buzz of a distant propeller is a constant reminder of imminent death' (2012). An al-Qaeda website noted the same: '[t]he harm is alarming, the matter is very grave. So many brave commanders have been snatched away by the hands of the enemies…by planes that are unheard, unseen and unknown' (cited in Williams 2010: 879). One Afghan villager echoed this concern. On the subject of American UAV strikes he said: 'we pray to Allah that we have American soldiers to kill. These bombs from the sky we cannot fight' (cited in Bearak 2001). UAVs, noted USAF Maj. Gen. Dunlap:

> [I]mpose the psychology of 'engagement dominance' on otherwise dogged adversaries. Death per se does not extinguish the will to fight in such opponents; rather, it is the hopelessness that arises from the inevitability of death from a source they cannot fight. (2007: 65)

For all the technological novelty of the UAV, the conceptual factors that underpin its emergence are as old as war itself. Belligerent powers have always sought to insulate the combatants of their own side, as much as reasonable, from the threat of reciprocal harm. What is new, however, is the degree to which UAVs intensify this mitigation of physical risk, particularly when utilized as the *sole* means of battlefield violence. The United States currently stands at the forefront of this process and is increasingly able to restructure

[18] Armed UAVs also rely on supporting infrastructure, including ground control installations, communications, and GPS, all of which may be susceptible to hostile state interference (Mayer 2015: 768–9).

the battlefield under conditions of radical asymmetry. The United States also stands at the centre of the ethical controversy that has emerged as a result of this imbalance.

The Ethical Challenge of UAV Violence

Like the two historical episodes of asymmetry examined in Chapters 5 and 6, the ethical challenge of UAV violence stems primarily from the perception that its operators kill in the absence of a sufficient degree of personal physical risk. After detailing the nature and severity of this contemporary challenge, this section will assess its potential for resolution. It is argued that just as with military sniping and manned aerial bombing, the ethical status of UAV pilots is likely to normalize.

UAVs and the End of Heroism?

Even within the hyper-technical and casualty-averse setting of modern war, the warrior ethos continues to hinge to a significant degree on the recognition that honour can be accrued on the battlefield only when fighters subject themselves to a sufficient degree of physical risk. The extent to which UAV pilots bypass this expectation stands as a principal factor in their contested ethical status. Sparrow exemplifies this opposition, arguing that the utilization of UAVs against those 'who have no opportunity to return fire' is 'inherently dishonourable' (2013: 98–9). 'Unmanned systems', argues Lee, 'create the first complete break in the ancient connection that defines warriors from their soldiery values' (2012: 13). For Enemark, this break has eroded the 'virtue of the warrior as a courageous risk-taker' (2013b).[19] Unsurprisingly, those targeted by such violence have voiced similar criticism. In 2011, former US Director of National Intelligence Dennis Blair acknowledged that:

> As the drone campaign wears on, hatred of America is increasing in Pakistan...Our reliance on high-tech strikes that pose no risk to our

[19] See also, Chamayou (2015: 101); Exum (cited in Mayer 2009); Riza (2013: 89); and Williams (cited in Kaldor and Mair 2011: 14.55).

> soldiers is bitterly resented in a country that cannot duplicate such feats of warfare without cost to its own troops. (2011: A21)[20]

Of greater significance is the extent to which this criticism is reflected within Western militaries themselves. The USAF was initially resistant to the adoption of UAVs. The basis for this, at least in part, was the tension between the riskless dimensions of this technology and the 'Top-Gun' masculine culture inherent to the institution's history (Singer, cited in Gusterson 2014: 194). Moreover, UAV operators have been derided by their fellow combatants as 'desk-jockeys' (Chapelle, cited in Woods 2015: 179) and 'chair-borne rangers' who realistically would 'only earn a Purple Heart for burning themselves on a Hot Pocket' (cited in Power 2013). Britain's Air Chief Marshall Sir Brian Burridge has gone even further, describing UAV 'hunter-killer' missions as 'virtue-less war' (cited in Mayer 2009).

In a 2006 survey of 400 American officers with aviation specialities, a majority of respondents rejected the notion that UAV operators would ever be recognized with the Distinguished Flying Cross (DFC), an award given for 'heroism or extraordinary achievement' in the air. The overwhelming reason for this rejection was 'that at least for some level of personal awards, human risk should always be a distinguishing criterion' (Fitzsimonds and Mahnken 2007: 101).

This sentiment was seemingly borne out in February 2014, when the Pentagon introduced a new honour termed the 'Distinguished Warfare Medal'. It applied to UAV operators and ranked above both the Purple Heart (awarded to those wounded or killed in duty) and the Bronze Star (awarded for heroic or meritorious achievement). Its purpose, according to then Secretary of Defence Leon Panetta, was to 'acknowledge achievements that directly impact on combat operations, but that do not involve acts of valour or physical risk that combat entails' (cited in Shanker 2013). Though Panetta was careful to qualify its scope away from active combat, the proposed medal nevertheless drew significant criticism. In response to the episode, the American Legion issued a statement regarding the need to recognize the 'fundamental difference between those who fight remotely, or via computer, and

[20] In addition to its ethical implications, the elimination of battlefield risk for one side carries with it a political challenge. British Group Capt. Clive Blount has argued that the use of lethal force at zero risk 'enables the insurgent to cast himself in the role of underdog and the West as a cowardly bully' (2011: 34–5). Before his death, Osama bin Laden claimed that the dependence of the United States on UAVs, 'conceal[ed] their most prominent point of weakness, which is the fear, cowardliness, and the absence of combat spirit among US soldiers' (2003: 12). See also, Fierke (2013: 209); McChrystal (cited in Alexander 2013b); Pryer (2013); and Wilkerson (cited in Tenfold 2016).

those fighting against an enemy who is trying to kill them' (ABC News 2013). Pete Hegseth, then CEO of Concerned Veterans of America, stated simply: 'heroism…is something that requires risk' (cited in Fox News 2013: 1:43). So great was the opposition from active and former members of the military, that Panetta's successor, Chuck Hagel, opted to cancel the new combat medal.[21]

On the basis of such criticism, one could be forgiven for concluding that the status of UAV pilots, at least in terms of their connection with the warrior ethos, is irreparable. What this overlooks, however, is the inherent adaptability of this ethical framework.

Ethics in an Age of Radical Asymmetry

Adherence to the explicit rules and standards of war is an increasingly powerful determinant of battlefield legitimacy. The ability of UAV pilots to align with this more responsibility-driven iteration of the warrior ethos is one way in which the ethical legitimacy of their violence can be sustained. UAV operator and squad commander Col. Eric Mathewson suggested as much in 2010, arguing: '[v]alour to me is not risking your life. Valour is doing what is right. Valour is about your motivations and the ends you seek' (cited in Enemark 2013a: 86). A 2016 study of the formation of ethical subjectivity among British Reaper crews highlighted the extent to which this shifting conception of the warrior ethos had been internalized by the operators themselves. Of the twenty-four Reaper pilots questioned, frequent reference was made to the perceived need to 'do the right thing', particularly in relation to the existing rules of engagement and laws of war (Lee 2014: 6).

This normalization of status is likely to be accelerated in situations in which UAVs work in conjunction with troops on the ground. Just as with military sniping and manned aerial bombing, UAV pilots operating in this context are able to draw their ethical legitimacy not from their own exposure to physical risk, but rather, their capacity to alleviate it for others. As one UAV operator stated:

> The air support that I am able to provide [troops on the ground]…directly improve[s] their rate of survival. If conducted ethically and effectively they

[21] Seemingly undeterred by this criticism, Sir Michael Fallon recently stated that the United Kingdom would conduct a review of how best to recognize the contribution of UAV operators, including the option for 'medallic recognition' (cited in BBC News 2017).

> also minimise the extent to which they and the enemy are subjected to that conflict, hopefully hastening its resolution. Whilst this may appear to be a lofty sentiment, the direct effect my actions have on improving a soldier's life in a conflict zone is the primary motivation. (cited in Lee 2014: 20)

This aspect of UAV violence is not lost on those that continue to engage with more traditional levels of physical risk. Col. Theodore Osowski, a ground-based unit commander in Kandahar, described UAVs in the following way: 'it's kind of like having God overhead, and lightning comes down in the form of hellfire' (cited in Mockenhaupt 2009). Other ground-based troops expressed similar gratitude for UAV pilots, referring to them as 'angels in the sky' (cited in Power 2013). This is not to argue that an ethical challenge does not exist. Alongside these statements of support are those provided earlier, that point to a profound ethical discomfort, grounded in the riskless capability of the UAV. This contested narrative of legitimacy has yet to be settled. What is suggested here is that, just as with the asymmetry-challenges of the past, there is strong potential for the gradual ethical normalization of this category of violence.[22]

The arguments presented in this section would seem to validate the view that UAV asymmetry is merely the latest in a long line of resolvable ethical tensions within the history of war. Problematically though, this is not the case. UAV violence, when undertaken to the exclusion of higher-risk alternatives, challenges the basic assumption of war as a contestation of structural reciprocal risk. This, it is argued, has the potential to destabilize the coherence of the moral justifications for lethal force, specifically in terms of the permissiveness of inter-combatant violence.

The Moral Challenge of UAV Violence

It is firstly worth clarifying that UAV strikes are but one of a number of moral and legal controversies to have arisen in the wake of America's ongoing War on Terror. From its genesis, the campaign was framed by the Bush

[22] This resolution would likely be hastened by greater awareness of the non-physical costs associated with the use of UAVs. Many have argued that UAV operators experience post-traumatic stress disorder (PTSD) at a rate that rivals, and potentially even surpasses, pilots of manned aircraft (Axe 2012; Duggan 2010: 1; Dunlap 2014: 126–7; Lee 2012: 13; Martin 2011; Singer 2010). Adding to this is the risk of 'moral injury'—trauma caused by a dissonance between a combatant's own ethical understanding of war and their actions during battle (Abé 2012; Kirkpatrick 2016; Shay 2003). A lack of physical risk does not automatically translate into an absence of sacrifice.

administration as a 'state of exception'; a conflict of such import, the circumvention of particular international norms, including those of Just War, could be rationalized (Ralph 2013: 1). Then US Attorney General, Alberto Gonzales, encapsulated this view when he dismissed parts of the Geneva Conventions as 'quaint' and 'obsolete' (cited in Roth 2017).

This logic underpinned a range of questionable actions taken by the United States during this period, including its designation of captured enemy fighters as 'unlawful combatants'. Among the critics of this approach was Barack Obama, who as President sought to justify UAV strikes using long-established frameworks of legal and Just War theory. In a 2013 speech, Obama defended the US UAV campaign as a 'Just War', citing a 'near-certainty standard' of no civilian deaths (2013). As this defence highlights, the *in bello* permissibility of the UAV is most often considered exclusively in terms of its adherence to the principles of discrimination and proportionality. UAV violence poses a moral challenge, it is commonly argued, only when it exacerbates the risk of injury and death to non-combatants. In contrast, its potential to call into question the moral justifications for targeting *the otherwise targetable* has received comparatively little attention. What this framing overlooks is the extent to which the coherence of the permissive moral standards of inter-combatant violence is grounded in an assumption of war as a site of structural reciprocal risk. UAV-exclusive violence challenges this assumption, and with it, the moral premise of the permissiveness itself. The following section examines this in relation to the three most dominant positions of Just War, beginning with contractarianism.

Radical Asymmetry and Just Conduct in War

The contractarian account of Just War presents the mutual right of combatants to kill as the outcome of an implicit contractual relationship between collective belligerents. Under this 'contract', the non-combatants of both sides are preserved from direct and deliberate violence. Combatants, in contrast, are entitled to kill, while at the same time targetable themselves to lethal force. The reciprocal application of this division renders it a 'procedurally fair and mutually beneficial norm' (Benbaji 2008: 487). Chapter 4 argued that the coherence of this norm hinges to a significant degree on the realistic capability of both sides to target those deemed liable to harm. In short, it demands the maintenance of structural reciprocal risk. This contract risks breakdown when the battlefield is restructured under conditions of unilateral violence.

Ryan observes that 'recent technological changes in warfare have altered the practice of war in ways that inherently undermine the conditions needed for conventions to operate as they traditionally have' (2012: 46). One of the most destabilizing of these technological changes has been the emergence of UAV-exclusive violence. A major benefit of both military sniping and manned aerial bombing was the extent to which they restricted the capability of those targeted to effectively retaliate on the battlefield. Yet in neither case was this threat-potential truly nullified. The substitution of human combatants for machine proxies goes well beyond these historical asymmetries, challenging the very functional premise of the contractarian justification for killing in war.

One obvious counter to this argument is the enormity of the moral crimes perpetrated by many of those targeted by American violence. As Statman writes: '[a]ccording to... [contractarianism]... nothing could be wrong with the use of... [targeted killing] against organizations like al-Qaeda which have no respect for the conventions of warfare' (2012: 106).[23] There is no question that the conduct of some of these non-state organizations, particularly their deliberate targeting of civilians, places considerable strain upon the adversarial bargain of contractarianism. Crucially though, these crimes do not negate the strain *also* applied by the refusal of the United States to expose its own forces to the reciprocal risk of the battlefield. 'It is difficult to imagine', suggests Lt. Col. Douglas Pryer, 'how anyone could feel that their enemies were justified in waging war against them via remote-controlled machines... if there were no way they could reply in kind... the situation seems fundamentally unfair, unjust, or unreciprocal' (2013: 19). UAV-exclusive violence challenges the assumption of structural reciprocity upon which the coherence of the contractarian account rests. In the absence of this coherence, the permissiveness of American violence loses much of its moral force. This assumption of structural reciprocity also applies to the conventionalist account of Just War; as does, by extension, the moral challenge of radically asymmetric violence.

Conventionalism grounds its justification for inter-combatant violence on the material threat, rather than moral culpability, of those targeted. Drawing upon the logic of self-defence, this position morally empowers combatants to target those on the battlefield that pose a mortal danger; i.e. other

[23] Statman further contends that even if weaker parties did adhere to the war convention, radically asymmetric UAV violence would not constitute a challenge to the contractarian justification for lethal force (2012: 106).

combatants. Some have argued that UAVs neutralize the physical risk of their operators to a degree that calls into question the coherence of the conventionalist justification for lethal force. According to Zehfuss:

> The use of weapons that put combatants out of their enemies' range invalidates one of the most common arguments for the permissibility of targeting enemy combatants, namely that they are in the business of harming the combatant doing the targeting. (2011: 555)

What Zehfuss insufficiently credits here, however, is the *collective* conception of 'risk' and 'threat' that underpins the conventionalist account. Recall that war is not only a violent contestation between individuals, but also collective belligerents. Combatants are, it is presumed, simultaneously an individual material threat and a constitutive element of the structural threat of their side. This presumption of collectivized threat morally justifies the targeting of combatants on the basis of *status*, rather than the narrower and less permissive metric of *conduct* (Zohar 1993: 615). The challenge to the conventionalist justification for lethal force is not, as Zehfuss suggests, inherent to the UAV. Rather, it is contingent on a level of military imbalance currently achievable only through the practice of UAV-*exclusive* violence. Within these specific conditions, the collective threat of the weaker side is mitigated to a degree that invalidates the moral justification for *status*-based targeting.

The obvious should first be acknowledged; even within conditions of radical asymmetry, an ongoing diffuse threat against the United States remains in the form of terrorism.[24] The extent of this enduring structural reciprocity, however, must be properly recognized. According to terrorism datasets, the likelihood of an American citizen being killed in a terrorist act is one in 3.5 million per year, in contrast to a one in 8,000 chance of being the victim of homicide (Mueller and Stewart 2012: 95–6). More recent studies rank the odds of death in a terrorist attack slightly above being killed in a tornado, and significantly less than being killed by the police (Mosher and Gould 2017).[25] Even if we factor in the generalized threat of terrorism, radically asymmetric UAV violence has mitigated the collective threat of those targeted to a degree

[24] The use of UAVs in Yemen has, at certain times, been justified on the basis that al-Qaeda in the Arabian Peninsula (AQAP) represents an ongoing threat to the West (Gusterson 2016: 19; Salisbury, cited in Higgins 2013).

[25] These comparisons can only be drawn so far. The potential psychological impact of terrorism on the general public is far greater than that of most natural calamities. For many, this is sufficient justification, at least politically, for more robust methods of deterrence.

far beyond prior warfare, including the asymmetric battlefields of both the First Gulf War and OAF. Lubell has argued that the full tally of terrorist activity directed against the United States over the past two decades, including the attacks of 11 September, still only amounts to 'isolated and sporadic' violence (2010: 117).

The argument here is that the totality of this threat is insufficient to morally justify the use of status-based targeting within conditions of UAV-exclusive violence.[26] This is enough to problematize the 'global war on terror' standard set by the Bush administration, which justified the lethal targeting of suspected terrorists wherever they were located on the basis of group membership (Daskal 2013: 1176). This does *not* mean that violence can no longer be exercised. Those targeted by the United States outside the active battlefield may very well be liable to direct and deliberate violence. Crucially though, this determination must be reached on the basis of the target's specific *conduct*. It is this individualist metric that is employed by Just War revisionists.

Revisionism draws no distinction between the moral rules that govern conduct in war and those applicable to any other situation. This position rejects, therefore, the idea of a 'moral equality' between combatants. Rather, violence may be directed only against those who sufficiently contribute to an *unjust* threat. This account of Just War also opposes *collectivized* determinations of liability, and by extension, status-based targeting. According to Lazar:

> Complicity is irrelevant to the ethics of killing in war. You cannot be liable to be harmed defensively unless you have *yourself* done something that contributed to a threat that can be averted by harming you. (2015: 132)

In more recent years, the policy of the United States, at least in relation to UAV-exclusive violence, has shifted to reflect this individualist account of enemy liability. In May 2013, then President Obama issued a 'presidential policy guideline' limiting the use of UAVs outside active war zones to targets that present a 'continuing, imminent threat to U.S. persons' (2013).[27] This followed on from a 2012 statement, in which Obama claimed that such strikes

[26] The problem of conceptualizing the enemy as a sufficiently 'unified threat' is compounded by the difficulty in establishing the precise allegiance of specific terrorists, as well as the differences that exist in motives, goals, and most importantly, *threat potential* between different terrorist organizations. For a legal perspective on this challenge, see Ohlin (2012: 64–5).

[27] This was reaffirmed in July 2016, when the Obama administration provided long-awaited statistics regarding the use of UAVs outside 'hot' battlefields (Shane 2016).

were restricted to threats that were 'serious and not speculative' (cited in The Economist 2012). In that same year, John Brennan stated publicly that:

> [W]hen considering lethal force [beyond 'hot' battlefields] we ask ourselves whether the individual poses a *significant* threat to U.S. interests...I am not referring to some hypothetical threat, the mere possibility that a member of al-Qaida might try to attack us at some point in the future.
>
> (2012, my emphasis)

Former Attorney General Eric Holder echoed this, suggesting that outside the active battlefield, strikes were directed only against 'specific senior operational leaders' (2012). According to former Secretary of State, John Kerry, UAV strikes were restricted to 'confirmed terrorist targets *at the highest level*' (cited in BBC News 2013, my emphasis).

Even within the context of radical asymmetry, the killing of high-level enemies may be morally justified through an individualist, *conduct*-based metric of liability. Such individuals, it is reasonable to assume, play an essential role in the planning, operationalization, and execution of terrorist attacks. They are causally connected to a *sufficient* degree to the 'continuing and imminent' threat faced by Americans, even if geographically and temporally dislocated from the threat itself (Montague 2012: 286–7; Tesón 2012: 411–13). The question this immediately raises, however, is at what point does this assumption of liability become *unreasonable*? How are we to reconcile conditions of radical asymmetry with the killing of those whose contribution to the threat in question is more tenuous?

The next section engages the actual empirics of UAV strikes. It illustrates that contrary to the assurances of the Obama administration, UAV-exclusive violence has been directed by the United States primarily against low-level enemies. It is argued that the threat represented by these individuals is insufficient to sustain a *presumption* of liability to harm.

Low-Level Targeting and the Conceptual Expansion of Threat

The United States designates liability to UAV targeting on the basis of both 'personality' and 'signature'. 'Personality' strikes are undertaken against specific individuals identified prior to targeting. 'Signature' strikes, in contrast, function on a 'pattern of life' analysis, targeting 'groups of men who bear certain signatures, or defining characteristics associated with terrorist activity, but whose identities aren't known' (Bowcott 2012: 19). One US counterterrorism official described these characteristics as 'actions over time [that make]...it obvious they are a threat' (Cloud 2010). Although the exact makeup of these 'signatures' remains unclear, evidence suggests they include actions such as planning attacks, transporting weapons (Entous, Gorman, and Barnes 2012), handling explosives (Cloud 2010), as well as being present in a compound or training camp for terrorism (Miller 2012). These more controversial strikes commenced in 2008 and have since been conducted in Pakistan, Yemen, and Somalia (Currier and Elliot 2013; Daskal 2013:1186; Kaag and Kreps 2014: 31–2).[28] In each of these regions, questions have been raised regarding the moral justifications for American violence.

A Reuters investigation conducted in 2010 reported that of the 500 individuals killed by UAV strikes in Pakistan since 2008, only fourteen were considered 'top tier militant targets', with twenty-five deemed 'mid-to-high level' (Entous 2010). Low-level al-Qaeda and Taliban fighters were targeted and killed at a rate twelve times greater than mid-to-high-level fighters (Entous 2010).[29] An evaluation of the Pakistan UAV programme conducted the following year by the Obama administration found that low-level militants remained the disproportionate focus of strikes (Bergen and Rowland 2012).

In Yemen the figures are similarly troubling. By 2014, CIA and JSOC strikes, mostly via UAV, had killed between 500 and 1,400 Yemeni targets (TBIJ 2016). Though initially restricted to al-Qaeda's senior leadership, targeting subsequently expanded to include local opposition to the authoritarian president, Ali Abdullah Saleh (Woods 2015: 205). The United States, argued Ryan Devereaux, was 'devoting tremendous resources to kill off a never-ending stream of nobodies' (2015). Even some State Department officials regarded the criteria used by the CIA for determining terrorist 'signatures' as 'too lax' (Becker and Shane 2012: A1).[30] The nature of these strikes suggests an over-permissive targeting criterion by the United States.

A crucial point to consider here is that these strikes would likely *not* be morally problematic—regardless of the individual threat level of the targeted fighter—if conducted in a more traditional war zone. Within such conditions,

[28] According to one US official, twice as many 'wanted terrorists' have been killed by the United States in 'signature' strikes than in 'personality' strikes (Entous, Gorman, and Barnes 2012).

[29] For similar figures, see Barrinha and Vinha (2015: 23); Brooks (2013); Brooks (2014); Human Rights Clinic (2012: 8); Landay (2013); Linden (2015: 173); and Telhami (2004: 16).

[30] In reference to Pakistani signature strikes, one former UAV operator claimed: 'the impression I got was that those people were arms runners and messengers' (Woods 2015: 21).

liability to harm may be justifiably established on the basis of *general presumptions*. Daskal writes that in an actively contested battlefield:

> [W]here troops on the ground are exposed to high levels of risk... those engaged in on-the-ground combat should not be required to hold their fire until they conduct a careful evaluation of the threat posed; such a rule would be potentially suicidal. (2013: 1198)

The permissiveness with which combatants may violently target one another on the battlefield is grounded in an assumption of reciprocal structural risk. Radically asymmetric violence undermines this assumption to a degree that calls the coherence of this permissiveness into question. Within such conditions, the *presumption* that low-level enemies are sufficiently liable for direct and deliberate targeting is no longer morally justifiable.

Chapter 4 suggested two problematic outcomes of this moral tension. The first was the potential killing of individual fighters who, as a consequence of the radically asymmetric conditions of the battlefield, were no longer liable to lethal targeting. Evidence of this can already be found in the frequency with which UAV strikes have been directed against low-level enemies. The second possible outcome would be a lowering of the liability threshold itself, to include increasingly tangential and speculative threats. Evidence of this second, and potentially more serious outcome can be found in the efforts by the Obama administration to reinterpret the concept of 'imminence' in the context of UAV targeting.[31] In 2013, then State Department legal advisor Harold Koh coined the somewhat paradoxical term 'elongated imminence':

> [W]hich he likened to 'Battered Spouse Syndrome.' If a husband demonstrated a consistent pattern of activity before beating his wife, it wasn't necessary to wait until the husband's hand was raised before the wife could act in self-defence. Similarly, terrorists wouldn't have to be boarding the plane with bombs before American commandos could take them out.
> (Klaidman 2012: 219–20)

The extent of this reinterpretation was further clarified in a leaked 2013 memo from the Justice Department, which outlined that for the United States, '[imminence] does not require... clear evidence that a specific attack on US

[31] Though much of this debate has been conducted in reference to the law, the focus here remains on its moral dimensions.

persons and interests will take place in the immediate future' (US DOJ 2013: 7). It added as further justification that 'al-Qaeda would engage in such attacks regularly to the extent it were able to do so', and that the 'U.S. government may not be aware of all al-Qaeda plots as they are developing and thus cannot be confident that none is about to occur' (US DOJ 2013: 8).

This expanded conception of imminence has contributed to an over-inclusive targeting criterion in relation to UAV strikes, particularly when such strikes are conducted in conditions of radical asymmetry. Whetham shares these concerns, arguing that an 'elongated' conception of imminence allows for the authorization of force on the basis of increasingly distant and questionable threats (2013: 78). Regarding this shift in threat perception, former Air Force lawyer Morris Davis stated: 'I'm thankful that my doctors don't use their definition of imminence when looking at imminent death. A head cold could be enough to pull the plug on you' (cited in Landay 2013).

This conceptual reinterpretation also increases the risk that such strikes will be justified not on the basis of present or future threats at all, but rather, as punishment for suspected involvement in past attacks (Whetham 2013: 77). To be clear, such individuals may of course be culpable for prior wrongful acts. Crucially though, what is under discussion here is not moral opprobrium, or even criminal liability to arrest and prosecution, but rather an immediate death-sentence via UAV missile. In order for such a severe and irreversible response to be morally justified within conditions of radical asymmetry, it is incumbent upon the United States to establish a clear causal connection between the target and relevant threat. Obama himself seemed to acknowledge this need in a 2013 speech, stating: 'not every collection of thugs that labels themselves al-Qaeda will pose a credible threat to the United States' (2013). The existing evidence, however, points to a discord between the rhetoric of the United States and its actual UAV targeting practices.

When engaging this issue, it is again worth reflecting on the qualitative distinction between UAV-exclusive violence and the asymmetry-challenges of the past. For all their impact on the battlefield, neither sniping nor manned aerial bombing imposed a degree of military imbalance sufficient to undermine the basic assumption of war as a site of pervasive, reciprocal, and *structural* risk. It remained possible in both cases, therefore, to morally defend the permissiveness of inter-combatant violence. Radical asymmetry fundamentally challenges this assumption of structural reciprocity, and with it, the tacit rationale of the permissiveness itself.

This break with the past is equally stark in terms of the implications of this challenge. This chapter concludes by arguing that beyond its potential to render less tenable the presumptions by which moral liability to harm is designated, UAV-exclusive violence signifies a fundamental shift in the *nature* of war. This shift is likely to have a damaging impact on the regulation of lethal force.

The Implications of Radical Asymmetry

Enemy Status and the Changing Nature of War

Chapter 2 argued that while radical asymmetry does not mark the *end* of war, it does represent a fundamental shift in its nature. Whereas warfare has traditionally been structured as an adversarial contestation, radical asymmetry more closely approximates a one-directional dispensation of violence. This change in the basic logic of hostilities has negatively influenced the battlefield conduct of the United States. In the context of UAV-exclusive violence, it has manifested as a loss of recognition for the enemy's humanity and moral value.

Before detailing this process, we should recognize that long before the emergence of UAVs, concern has existed regarding the potential of military technology—particularly advances in warfare that extend the physical distance between opposing combatants—to conclude war as a *human* endeavour. In 1944, German sociologist and philosopher Theodor Adorno made precisely this claim in relation to Germany's V-weapons, arguing that they had fundamentally altered war as an inter-subjective experience. The introduction of these 'robot bombs', Adorno claimed, was but one aspect of a new and dispassionate mode of war fighting: 'of civil engineering and blasting operations undertaken with immeasurably intensified vehemence, also of "fumigation", insect-extermination on a terrestrial scale...' (2005: 60).

Radically asymmetric violence re-invokes these fears by calling into question our very conception of what constitutes war. In certain contemporary contexts: 'soldiers need no longer go to war expecting to die, but only to kill' (Asad 2007: 202). This unmooring of battlefield violence from physical risk 'destabilises the conventional understanding of war as an activity in which human dying and killing are exchanged' (Asad 2007: 202).[32] When this

[32] The mitigation of battlefield risk by the United States also has significant political implications, even when it falls below the standard of radical asymmetry. The Obama administration cited an absence of physical risk as a principal factor in its justification for military action against Libya in 2011 without the congressional approval required by the War Powers Act of 1973. In a 32-page report, the administration argued that: 'U.S. operations do not involve sustained fighting or active exchanges of fire with hostile forces, nor do they involve the presence of U.S. ground troops, U.S. casualties or *a*

exchange is absent and the battlefield restructured under conditions of unilateral violence, the conceptual and practical distinction between war fighting and extra-judicial punishment weakens.[33]

This is particularly true in the context of American UAV violence, given the selective use of a law enforcement paradigm throughout the War on Terror. President Bush himself stated in a 2003 speech that one of the principal advantages of the UAV was its capability to deliver 'sudden justice' to the enemies of the United States (2003). A year earlier, then Defence Secretary Donald Rumsfeld sent a memo titled 'Manhunts', in which he asked, 'how do we organise the Department of Defence for manhunts' (cited in Naylor 2015). UAVs were considered an essential component of this emerging strategy. Chamayou, in his book, *Manhunts: A Philosophical History*, defined war-as-hunting, or *cynegetic war*, as follows:

> (1) It does not take the form of a direct confrontation, but a process of tracking down; (2) the power relationship is marked by a radical dissymmetry in weapons; (3) its structure is not that of a duel: a third term is mobilized as a mediation; (4) the enemy is not recognized as such, that is, as an equal – he is only a prey; (5) use is made of nonnoble means related to policing or hunting rather than to the classic military register. (2012: 73)

Juxtapose this against an account provided by *The New York Times*, detailing the process by which the Obama administration determined UAV targets:

> It is the strangest of bureaucratic rituals . . . Every week or so, more than 100 members of the government's sprawling national security apparatus gather, by secure video teleconference, to pore over terrorist suspects' biographies and recommend to the president who should be the next to die.
>
> (Becker and Shane 2012: A1)[34]

Obama himself acknowledged that the UAV 'looks like a pretty antiseptic way of disposing of enemies' (cited in Chait 2016). Importantly though, the concern here goes beyond a question of aesthetics. The capability of UAVs to

serious threat thereof, or any significant chance of escalation into a conflict characterized by those factors' (Macmanus and King 2011: 25, my emphasis).

[33] See also, Clark (2015: 131–2); Dunn (2013: 1238); Enemark (2013a: 4–6); Gross (2010: 268); Riza (2013: 88); and Schell (2011).

[34] Former Shin Bet chief Ami Ayalon provided a similar account of Israel's targeted killing programme, describing the target selection as a 'kind of conveyor belt' (cited in Woods 2015: 168).

enable the delivery of lethal force with impunity pushes our understanding of such violence beyond any conceivable notion of 'fighting', a term that presupposes some degree of mutual contestation. Rather, what we are presented with is a form of technological predation; of pest-control—a 'gardener's vision of war':

> Weeding out is a creative, not a destructive activity… All visions of society-as-garden define parts of the social habitat as human weeds. Like all other weeds, they must be segregated, contained, prevented from spreading, removed and kept outside the society boundaries; if all these means prove insufficient, they must be killed. (Bauman 1989: 92)

The unilateral nature of UAV predation has eroded the moral relationship between the United States and those it targets. This erosion is compounded by a growing reliance on meta-data analysis. In many cases, individuals are killed not on the basis of their own threatening conduct at all, but rather their proximity to a suspicious mobile phone (Scahill and Greenwald 2014). The algorithmic nature of UAV signature strikes, writes Williams, 'precludes reciprocity', reducing those targeted to mere 'data streams' (2015: 105). According to one American intelligence source, the internal view of the special operations community towards those hunted by UAVs:

> [Is] 'they have no rights. They have no dignity. They have no humanity to themselves. They're just a "selector" to an analyst. You eventually get to a point in the target's life cycle that you are following them, you don't even refer to them by their actual name.' This practice, he said, contributes to 'dehumanising the people before you've even encountered the moral question of "is this a legitimate kill or not?"' (cited in Scahill 2015)

Evidence of this dehumanizing effect can also be found among UAV operators themselves. Former senior airman with the USAF Michael Hass flew UAVs for six years. He recounts during this time phrases such as 'cutting the grass before it grows out of control', and 'pulling the weeds before they overrun the lawn' to describe the use of UAVs against enemy insurgents (cited in Pilkington 2015).[35] Hass went on to say: '[e]ver step on ants and never give it

[35] For more detail on the dehumanizing impact of this asymmetry, see Chamayou (2015: 165–6); Clark (2015: 116); Gholiagha (2015); Muñoz-Rojas and Frésard (2004: 199); Welsh (2015: 43); Wilcox (2015: 151); and Zenko (2013b).

another thought? That's what you are made to think of the targets – as just black blobs on a screen' (cited in Pilkington 2015). Some have seized upon these admissions as evidence that the radically reduced physical risk of the UAV creates a 'PlayStation mentality'—a psychological disconnectedness among operators that loosens inhibitions on violence (Alston 2010: 25; Cole, Dobbing, and Hailwood 2010: 1–20; Finkelstein 2012: 174; Olsthoorn 2011: 126). This claim, however, has received considerable pushback from those that stress the intimacy and psychological strain inherent to the use of UAVs (Boyle 2015: 106; Gersten, cited in McCloskey 2009; Lee 2012: 15; Mason, cited in Woods 2015: 174).[36]

The most troubling implications of this dehumanization lie not at the level of individual combatants, but rather within the broader UAV policy of the United States itself. The ease with which such violence can be administered has incentivized the adoption of targeting practices that drift beyond the morally questionable, into direct violation. According to some reports, UAV 'signatures' have at times included being a military-age male in an area of known terrorist activity (Becker and Shane 2012: A1; Stanford Law School: 2012: 33); and 'consorting with known militants' (Filkins 2011). This approach, argued a number of counterterrorism officials, 'is one of simple logic: people in an area of known terrorist activity…are probably up to no good' (Becker and Shane 2012: A1). Evidence also exists of the US military retroactively classifying victims of collateral damage as 'enemies killed in action' (Scahill 2015). Perhaps the most troubling of these allegations relates to the practice of 'double tap' strikes—using those injured in initial strikes as bait, and then targeting a second strike on those that arrive to offer assistance (Casey-Maslen 2014: 397; Gholiagha 2015: 140; Kaag and Kreps 2014: 32–3).

The objective of this book was to determine the extent to which radically asymmetric violence challenges the moral justifications for inter-combatant violence. Its principal focus has been on questions of rule *coherence*, rather than *compliance*. With these examples, however, we witness the potential of

[36] Col. Pete Gersten, commander of the 342nd Air Expeditionary Wing at Creech, stated: '[a] lot of people downplay it, say "you're 8000 miles away. What's the big deal?" But it's not really 8000 miles away, it's 18 inches away…We're closer in a majority of ways than we've ever been as a service' (cited in McCloskey 2009). Another UAV operator rejected the notion of a 'PlayStation mentality' in even stronger terms: 'I used to fly my own air missions…I dropped bombs, hit my target load, but had no idea who I hit. Here I can look at their faces. I watch them for hours, see these guys playing with their kids and wives. When I get them alone, I have no compunction about blowing them to bits, but I wouldn't touch them with civilians around. After the strike, I see the bodies being carried out of the house. I see the women weeping and in positions of mourning. That's not PlayStation; that's real. My job is to watch after the strike too. I count the bodies and watch the funerals. I don't let others clean up the mess' (cited in Klaidman 2012: 217). See also, Thompson (cited in Cole 2017: 13–14).

radical asymmetry to more directly challenge the principles of battlefield restraint.

Evidence of this behavioural decline can also be found in the growing influence of politics on targeting selection. According to a number of sources, in the wake of the interrogation scandal of the Bush years, US officials have favoured UAV strikes as a more politically palatable alternative to capture (Anderson 2009; Becker and Shane 2012: A1; Hirsh 2011; Mazzetti 2013: 247). This shift in the grounds of acceptable force, from a morally justifiable standard of 'defensive' and 'necessary', to merely 'preferential', highlights the degree to which radical asymmetry strains, and in rarer cases, motivates outright violation of, the existing rules and standards of war.

To fully appreciate the problematic implications of this violence, we should remind ourselves of the domestic political context within which the UAV programme of the United States now operates. Between 2009 and 2017, the morally troubling aspects of this violence manifested under the guidance and overall supervision of President Obama, a constitutional law professor and determined critic of the military adventurism and overreach of the Bush years. There is now a new President, one who during the campaign at least openly advocated for a return of torture (Johnson 2016), as well as the deliberate violation of non-combatant immunity by American forces:

> And the other thing with the terrorists is you have to take out their families…they care about their lives, don't kid yourself. When they say they don't care about their lives, you have to take out their families.
>
> (Trump, cited in LoBianco 2015)

At the time of writing we are over two years into this new administration. President Trump has so far maintained the 'near-certainty' standard of no civilian deaths outside the official war zones of the United States in Iraq, Afghanistan, and Syria (Adams and Goodman 2018). At the same time, however, the administration has relaxed a number of other rules established by Obama to curb the problematic excesses of UAV violence. US intelligence officials will no longer be required to publicly disclose how many civilians have been killed in covert air strikes (Purkiss 2019). The Trump administration has also ended the requirement that beyond the 'hot' battlefield, strikes be undertaken only against those that pose a 'continuing, imminent threat to U.S. persons' (cited in Hartig 2017). Strikes are now formally permissible against 'foot-soldier jihadists with no special skills or leadership roles' (Savage and Schmitt 2017). This chapter criticized the Obama administration for its overly expansive interpretation of 'imminence'. That the term was retained at all, however, suggested at least a partial commitment by American policymakers to a more restrictive metric of targeting. The complete removal of this standard represents a troubling step-back in the regulation of this increasingly riskless mode of violence.

When assessing the implications of radical asymmetry, it is important to consider not only the conduct of the United States but also that of its opponents. This chapter concludes by arguing that the severity of asymmetry imposed through UAV-exclusive violence has incentivized those targeted to intensify attacks against the non-combatants of the United States.

UAVs and the Redirection of Risk

The use of UAVs in the absence of a significant ground presence conveys upon the United States an asymmetric advantage beyond anything witnessed in the history of war. Yet in achieving this, the United States has motivated its opposition to consider alternative targets. For those undeterred by moral opprobrium or legal penalty, this consideration has included non-combatants in the United States and elsewhere. American marine Maj. Trent Gibson suggested that:

> Attempts to armorize our force against all potential enemy threats... shifts the 'burden of risk' from a casualty-averse military force onto the populace. In doing so, we have lifted that burden from our own shoulders and placed it squarely upon those who do not possess the material resources to bear it – the civilian populace. (2009: 21)

Though the civilians Gibson refers to here are Afghans, rather than Americans, his warning is just as applicable to the domestic context. As already argued, the danger presented by terrorism is nowhere close to the level of threat that structures the traditional battlefield. Nevertheless, the terrorist threat that does emerge—or intensify—as a response to radical asymmetry presents an acute challenge.

One of most important tasks of the Just War Tradition is the creation and maintenance of a stark demarcation between the zone of war and zone of peace. UAVs, particularly when utilized to the exclusion of physically vulnerable ground forces, leave those targeted with few options for legitimate retaliation. The inevitable attempt by the weaker side to circumvent this dilemma has the potential to increase the burden of risk for the very citizens upon whose behalf the American military fights.

The imposition by one side of radical asymmetry destabilizes more than the moral frameworks that justify killing in war. This same imbalance forces us to reassess one of the most basic functions of the military, the protection of its citizens from violent harm. Radical asymmetry, if obtained at the expense of an enhanced civilian threat, compromises 'the traditional social division of danger', in which combatants are physically exposed and civilians physically insulated from harm (Chamayou 2015: 77). The upending of this social division should motivate reflection, not only on the changing nature of war, but also civil–military relations more broadly.

Conclusion

Within war there exist a range of weapons and methods that are considered so far outside the range of acceptability as to be *malum in se*, i.e. evil in themselves. The concept of *malum in se* has historically been invoked in reference to both nuclear and chemical weapons, as well as battlefield practices such as mass rape and genocide (Dige 2012). UAVs do not currently, nor are ever likely to, fall within this most reviled of categories. The moral challenge outlined within this chapter is not inherent to the technology. It is one of tension and strain, rather than straightforward violation. Specifically, it is argued that UAVs, when utilized to the exclusion of a meaningful ground presence, impose upon the battlefield an asymmetry of risk so profound as to call into question the permissiveness of American violence.

A central claim of this book is that the coherence of the moral justifications for lethal force is grounded in an assumption of war as a site of reciprocal structural risk between belligerents. It is this same structural reciprocity that stands as the critical point of delineation between the historical and contemporary challenge of asymmetry. The endurance of inter-belligerent risk in the context of both military sniping and manned aerial bombing limited the scope and resolution of both asymmetry-challenges to the warrior ethos. The challenge of UAV-exclusive violence includes, but extends beyond this ethical framework, to impact the Just War Tradition itself. This was illustrated in relation to the three most dominant accounts of Just War—contractarianism, conventionalism, and revisionism. It was argued that within each, a tension could be found between the radical asymmetry of UAV-exclusive violence and the assumptions that informed its permissiveness. This moral tension is most acute when violence is directed by the United States against low-level enemies.

This chapter concluded with an examination of the broader implications of radically asymmetric violence for the regulation of lethal force. UAV-exclusive violence has distorted the conceptual and practical distinction between war and judicial sanction to such a degree as to erode the moral relationship between the United States and its enemies. This same military imbalance has incentivized those targeted to intensify attacks against non-combatants. Both these outcomes place further strain on the rules and standards of the battlefield.

Conclusion

One of the most enduring representations of war throughout history has been the game of chess.[1] Like war, chess is innately adversarial, with both sides applying 'force' to overcome the other. Opponents in chess are distinctly uniformed, identically configured and equally empowered, engaging one another on a clearly defined battlefield, and in accordance with a commonly agreed set of rules (Rodin 2006: 153). As a result of these conditions, the 'combat' of chess is defined by a perfect symmetry of risk between enemies. Yet, as this description makes obvious, the game of chess is an abstraction that corresponds only with the most 'ideal-type' of war. In reality, battle has never fully embodied the symmetry of risk implicit in the chessboard image. Adversaries are governed instead by the opposite motivation. War 'is about winning, which means killing the enemy on the most unequal terms possible' (Junger 2010: 140).

Reinforcing this preference for asymmetry is the basic obligation of commanders and policymakers to insulate the military personnel under their charge from avoidable harm. Largely gone are the days when combatants were likened to cannon fodder and their lives expended without meaningful consideration. For both moral and political reasons, casualty aversion is an increasingly rigid constraint on the actions of belligerent parties, particularly liberal democratic states. Leaders who fail to utilize means and methods of warfare that might lessen the physical vulnerability of their side may be rightfully accused of callousness and negligence. Walzer argues that soldiers have the right to hold commanders accountable 'for every sort of omission, evasion, carelessness, and recklessness that endangers their lives' (2004b: 24).[2] In this sense, the mitigation of risk in battle can be understood as much as a duty, than a right among states.

[1] Military strategists have long analogized war to chess. During the American Civil War, for example, Robert E. Lee planned for the South to seize Washington in response to the potential capture of Richmond, declaring 'we would swap queens' (cited in Charnwood 1996: 302). Liddell Hart also drew parallels between the movement of knights and queens, and air-mobility and tank-mobility respectively (1954: 358).

[2] For examples of this accountability in the context of contemporary counter-insurgency, see Peters (2009); and Will (2010).

Asymmetric Killing: Risk Avoidance, Just War, and the Warrior Ethos. Neil C. Renic, Oxford University Press (2020).

DOI: 10.1093/oso/9780198851462.001.0001

This drive towards asymmetry, however, contains a contradictory element. Put simply, we seem more willing to countenance the *pursuit* of riskless killing in battle than its actual obtainment. As far back as the invention of the bow, opposition has existed to emerging modes of military violence that nullify the ability of those targeted to effectively retaliate. At various points in the history of war, reciprocal risk between opponents has been distorted to such a degree as to call into question the very legitimacy of such violence.

This book has examined the latest and most extreme iteration of this battlefield phenomenon—radically asymmetric violence: the sustained and theatre-wide capability of one side to kill in the near-complete absence of physical risk. This study was undertaken in order to determine the extent to which these conditions challenge the moral frameworks that justify killing in war. This book further considered the implications of such a challenge for the regulation of lethal force. In its response to both these questions, the history of war has been drawn upon extensively. First, to clarify the precise role of reciprocal risk within the frameworks themselves; and second, to assess whether there exists a *morally significant* distinction between the radical asymmetry of today and the now resolved asymmetry-challenges of the past. This book has argued that such a distinction does indeed exist. Beyond its more resolvable tension with battlefield ethics, radically asymmetric violence represents a fundamental challenge to the very coherence of the moral justifications for inter-combatant violence.

The Dilemma of Radical Asymmetry

Nobel Peace Prize winner Jody Williams has been a forceful and enduring opponent of the use of armed UAVs in war. She argues that 'when you can sit in Nevada and kill someone 7000 miles away...there's something unethical or immoral about that' (cited in Kaldor and Mair 2011: 14.55). This statement reflects the position of many who struggle to align the riskless aspect of UAV violence with their intuitive sense of what is 'right'. Moral intuition, however, though perhaps compelling, can only take us so far before it must be reinforced by a more theoretically grounded critique. Such an account is especially necessary if we are to properly reckon with the significance and implications of the historically unprecedented levels of asymmetry we are confronting on the battlefield today. Why are so many discomforted by modes of violence that severely distort the reciprocal risk between enemies in war? Is it the lack of physical *risk* sustained by those who kill that marks such violence as problematic, or the presumptive absence of physical *threat* projected by those targeted? These aspects of asymmetric violence, though frequently interconnected, are distinct.

Chapter 3 argued that in the case of the warrior ethos, it is predominantly the former—the non-risk of those who kill. This battlefield code of conduct encompasses a set of shared values and beliefs, developed and maintained by the participants of war to preserve their own ethical status. In determining this status, the warrior ethos has long-valorized the willingness of fighters to endure and overcome the physical risk inherent to battle. It has at the same time inculcated a disdain among its adherents for the use of violence in the absence of a sufficient degree of physical vulnerability.

The nature of this disdain was revealed in Chapters 5 and 6, which examined the historical asymmetry-challenges of military sniping and manned aerial bombing respectively. Upon their introduction, both modes of violence were widely viewed to insulate combatants from the physical risk of battle to a degree that imperilled their ethical standing. So intense was this opposition in the case of military sniping that when captured, snipers were often denied prisoner of war rights and instead hanged as criminals. A similar sentiment can be found in the historical failure of aerial bombing to attract the same level of veneration as aerial fighting. Crucially though, in neither case was this opposition permanent. Instead, we witness a gradual ethical normalization of both modes of violence. A key component in the resolution of these contested narratives of legitimacy has been the increasing importance of combat responsibility within the warrior ethos itself.

In light of this history, one could be forgiven for anticipating an equally straightforward resolution of the contemporary challenge of radical asymmetry. Indeed, in the context of UAV violence, evidence can already be found of the emergence of a counter-narrative of legitimacy, grounded in the professionalism and rule-conformity of UAV operators. This analogy between historical and contemporary episodes has helped reinforce an understandable consensus within much of the literature that radical asymmetry presents no *new* challenge to the justifications for killing in war. Suggestions to the contrary are typically framed as mere nostalgia for a standard of 'fair-fighting' that never truly aligned with the realities of battle.

This approach is mistaken. The moral challenge of radical asymmetry stems not from an idealized notion of 'equal risk', but rather, in the capacity of this mode of violence to nullify the *threat* of the weaker side. In contrast to the warrior ethos, the primary focus of the *in bello* framework of Just War are the potential targets, rather than agents of violence. One of the most basic and important functions of this tradition is to delineate the precise conditions by which individuals forfeit their right to be free of direct and deliberating targeting. Risk-asymmetry is relevant to the moral permissibility of killing in war only to the extent that it calls this forfeiture into question.

This was examined in greater detail in Chapter 4, which evaluated the three most influential accounts of Just War today—conventionalism, revisionism, and contractarianism. Though they differ to a considerable extent, all three endorse a highly permissive use of lethal force against combatants in war, at least in contrast to the type and degree of violence that would be permissible in most other circumstances. Combatants retain their liability to harm, even when posing no specific material threat when targeted. As this clarifies, reciprocal risk plays no *explicit* role in the moral justification for inter-combatant violence. A key claim of this book, however, is that the *coherence* of this permissiveness is ultimately grounded in, and contingent on, an assumption of war as a site of reciprocal *structural* risk. The liability of combatants hinges on their role as a constitutive element of the collective threat of their side.

It is at this collective, *inter-belligerent* level that the critical distinction between historical and contemporary asymmetry can be located. Neither military sniping nor manned aerial bombing sufficiently strained the assumption of reciprocal structural risk that underpins the permissiveness of inter-combatant violence. Within radical asymmetry, however, this assumption unravels, and with it, the underlying moral basis for the permissiveness itself. Within conventionalism, radical asymmetry undermines the image of the weaker side as a significant, unified threat, and by extension, the moral premise of *status*-based liability. Within revisionist accounts, it calls into question the presumption that enemy combatants, particularly those of a lower-level, are sufficiently causally connected to the unjust *collective* threat of their side. Even if one applies a more consequentialist approach, radically asymmetric violence remains a challenge. By undermining the long-standing assumption of war as a practice in which the militaries of both sides face harm, radical asymmetry destabilizes the implicit adversarial bargain at the heart of contractarianism.

Chapter 7 reinforced these findings empirically, detailing the challenge represented by UAV-exclusive violence to the moral justifications for killing in war. UAVs, when utilized to the exclusion of a significant ground presence, restructure the battlefield from a site of two-directional to one-directional force. Within such conditions, the permissive standards of inter-combatant violence that would otherwise apply become increasingly difficult to morally defend. This is not to suggest that radical asymmetry *by itself* invalidates the moral permissibility of UAV-exclusive violence. What has been argued is that in the absence of reciprocal structural risk, the capacity of existing moral rules to properly regulate such violence is diminished. This challenge is arguably at its most acute when UAV-exclusive violence is directed against low-level targets. Within conditions of radical asymmetry, it is increasingly difficult to maintain the *presumption* that such individuals are morally liable to summary execution.

Too many within the literature dismiss the significance of radically asymmetric violence, on account of its failure to directly violate the permissive moral rules that govern inter-combatant violence. Yet what such a dismissal fails to properly consider is the assumption of reciprocal structural risk that informs this permissiveness. The moral significance of radical asymmetry is grounded in its challenge to rule-*coherence*, rather than compliance. Conformity with the moral rules of war matters only for as long as the rules themselves remain effective signifiers of appropriate conduct in battle. Radical asymmetry threatens to undo this, by undermining the *presumptive* liability of targeted enemies. What we are then left with is a diminished moral framework, one less capable of delineating the reach and limits of battlefield lethality.

In order to establish the wider implications of this challenge for the regulation of lethal force, this book asked an additional question: to what extent has radical asymmetry transformed the *nature* of war? As we move closer to conditions of unilateral violence, we witness a fundamental change in the basic logic of hostilities, from an adversarial contestation to something approximating judicial sanction. The perception of the weaker party likewise shifts, from an opponent to be fought and overcome, to one that must be hunted and *done away with*. In the context of UAV-exclusive violence, this transformation in the dynamic between the United States and its enemies has contributed to a dehumanization of the latter. In some cases, this has led to a more severe decline in the targeting practices of the United States, from the morally questionable to the morally indefensible.

Radical asymmetry incentivizes an even greater rejection of restraint among those targeted. The intensity and entrenched nature of this military imbalance makes a redirection of violence more likely by the weaker side—away from the now insulated military forces of the United States and towards its still-vulnerable non-combatants. Radical asymmetry is not the sole explanation for this redirection. Such imbalance is, however, an important factor in its acceleration.

There is a specific reason why 'dilemma', rather than 'problem', best characterizes the emergence of radical asymmetry. War is replete with problems—difficulties in need of resolution. Indeed, most war crimes fit this description. All too frequently though, those who wage war are also required to navigate a range of moral dilemmas, situations in which no desirable outcome is obvious. What is under discussion is not how best to prevent an objectively wrongful act committed by one party on the battlefield. Participants in war have long sought to neutralize both the threat-capacity of the other side and the risk exposure to their own. The drive towards risklessness in battle is *logical* as dictated by the exigencies of war, and *laudable* as dictated by the moral and political demands of force protection.[3] This same drive, however, is moving us ever closer to a mode of violence that strains the basic assumptions upon which our moral justifications for killing in war rest.

While the focus of this book has been on the moral, rather than legal conventions of war, neither can be fully considered in isolation of the other. The following section will provide a brief outline of this interconnectedness. It is argued that radical asymmetry generates a problematic ambiguity, not only within the moral right to kill itself, but also in the *relationship* between this right and the legal guidelines of war.

The Coming Apart of Morality and Law

'Law', writes Kennedy, 'now shapes the institutional, logistical and physical landscape of war' (2012: 161). Smith agrees, arguing that 'never has the conduct of war been so legalistic' (2002: 356). Given this significant and growing influence, it would seem reasonable to conclude that it is the laws of war, rather than the warrior ethos or Just War Tradition, that we should first consult when evaluating the potential challenge of radically asymmetric violence. At first glance, this shift in focus would appear to resolve much, if not all, of the tensions associated with this issue. As Blum observes, within mainstream interpretations of the laws of war, there exists a 'general acceptance…of the near-absolute license to kill all combatants' (2010: 72). For as long as these laws remain applicable, this licence endures, no matter the degree of military imbalance. Before relegating the challenge of radical asymmetry to a marginal and exclusively moral concern, however, the critical and necessary

[3] Both these claims are generally true only. Within counter-insurgency warfare, winning the 'hearts and minds' of a population may require a dialling back of force protection. Now-retired US Army Gen. Stanley McChrystal famously disdained the use of body armour in an attempt to win over the Afghan public through a less militarized approach (Thomas 2009). See also, Pfaff (2011).

interconnectedness between the moral and legal frameworks of war should be recognized.[4]

This relationship was touched upon in the introductory chapter of this book, which presented the warrior ethos and Just War Tradition as two of the most important intellectual resources in the regulation of war. This status derives firstly from the role of each as an explicit guideline for appropriate conduct in battle. In addition, both have significantly influenced the formation of the positive and customary laws of war. Johnson describes the connection in war between moral values and law as 'an intimate and inseparable relationship' (1981: 71). This connection is ongoing, despite the consolidation of the latter as the dominant (and at times contradictory) means by which contemporary battles are adjudicated. It is to this relationship that Kels appeals when exploring the role of mutual risk in the laws of war. Kels first acknowledges that nowhere within this framework does the legality of inter-combatant violence hinge on the imposition of reciprocal physical risk. He goes on, however, to ask an important question: 'to what extent does the internal logic of [the laws of war] as a *coherent legal regime* rest upon a generalised notion of individual self-defence?' (Kels 2014, my emphasis). The logical follow-up to this question is how long can this legal regime *remain* coherent when self-defence has been so thoroughly transcended by one side?

Moral judgement retains an essential role in the laws of war, functioning as an ongoing threshold test for their acceptability. According to Garraway, just as morality without law is an aspiration, 'the law of war without ethics is an empty shell' (2014: 88). Shue holds a similar position. When we do not possess the 'morally best laws of war', Shue argues, 'the actual legal provisions ought to be changed' (2008: 90).[5] To ignore this is to risk a moral hollowing out of the laws of war. To better understand such a risk, the history of war can again be drawn upon. During the 1999 Kosovo bombing, NATO forces, including the United States, operated under strict rules of engagement. This was reinforced by a heavy presence of military lawyers, who were integrated into virtually every phase of the conflict:

> Military Lawyers... contributed assessments of the standard Geneva Convention questions for each target: was the objective military; were the

[4] The view expressed in this section aligns with Ronald Dworkin's position, that normative argument is inherent to legal reasoning. This is in opposition to the positivist separation of law and morality championed most famously by H. L. A. Hart. See Shapiro for a general overview of the 'Hart–Dworkin' debate (2007).

[5] Shue concedes that such change may be limited by what is practically achievable.

> means selected proportional to the objective; and what were the risks of damage to civilians. The texts of the Geneva Conventions themselves were available on screen. (Ignatieff 2000: 197)[6]

There were many, however, who questioned whether this strict technical compliance with the law came at the expense of deeper moral commitments. These concerns were underscored by the controversial NATO practice of dual-use strikes—attacks against objects that simultaneously served a civilian and military function. Critics pointed out that while dual-use attacks were legal, at least according to a permissive reading of the laws of war, they undermined a more fundamental obligation to the principle of discrimination. In doing so, they challenged the 'coherence of the LOAC [laws of armed conflict] as a whole' (Shue and Wippman 2002: 559–60).[7] This book makes a similar claim in regard to the emergence of radical asymmetry. The fact that such violence is legal does not resolve this challenge, but rather, points to a deeper one; specifically, the potential of profound discord between what is lawful on the battlefield and what is moral.

The immediate question this raises is what, if anything, can be done to narrow this divide between morality and law? One option is to tighten the legal restrictions that apply to radically asymmetric inter-combatant violence. Such a move would likely find support among those who already criticize the laws of war as overly permissive. Walker has suggested that the current laws of war are wrong to permit the use of lethal force against 'non-threatening' fighters (2012: 417). Blum has criticized *status*-based targeting on similar grounds, arguing that it fails to sufficiently protect physically defenceless combatants (2010).[8] Blum goes on to offer two amendments to the current doctrine of inter-combatant violence. The first of these:

> [C]alls for the relaxation of the status-based determination by supplementing it with an obligation to assess the individual threat emanating from any particular human target. This would require an amendment to the principle of distinction, from one separating combatants from civilians to one distinguishing threatening combatants from unthreatening ones... This amendment would essentially operate as the mirror-image of the doctrine of

[6] For more detail on the strengthening role of lawyers in contemporary armed conflict, see Coe and Schmitt (1997: 49); Dunlap (2001: 293); and Kramer and Schmitt (2008: 1407).

[7] See also, Farrell (2005: 162); Kaempf (2009); E. Schmitt (1999); and Shue (2016).

[8] See also, Alexander (1976: 413); Goodman (2013); Kasher (2014); Kasher and Yadlin (2005); and Nagel (1972: 140).

> civilian immunity…combatants [are] presumptively dangerous, unless they pose no or marginal threat. (2010: 108)

Blum's second proposal is for warring belligerents to incorporate a 'least-harmful-means test' into their military necessity calculations, mandating methods short of lethal force wherever feasible (2010: 74). Blum then lists a range of geographical and temporal factors that might inform the designation of liability to harm within this more restrictive framework.

The most obvious criticism to these proposals is that Blum overcorrects, swinging the legal pendulum from over-permissive to over-stringent. Indeed, the position of the preceding chapters is that the assumptions that underpin the existing moral and legal justifications for inter-combatant violence generally hold within the traditional structure of war. The suggestion, therefore, is that these restrictive measures apply only to conditions of radical asymmetry. In the context of UAV-exclusive violence, this would translate into a tightening of the targeting criteria of the United States. Specifically, it would severely restrict the use of 'signature' strikes outside 'hot' battlefields, particularly those directed against presumptively low-level targets. Such individuals could be justifiably harmed, but only in situations in which their specific conduct was determined to constitute, or sufficiently contribute to, an unambiguous and significant material threat.

An adjustment of the laws of war is unlikely to satisfy those who already contest the applicability of this framework outside the traditional battlefield. The basis for this contestation is the legal concept of 'intensity'. The ICRC provides the following criteria for the existence of a non-international armed conflict (NIAC):

> …*protracted armed confrontations* occurring between governmental forces and the forces of one or more armed groups, or between such groups arising on the territory of a State [party to the Geneva Conventions]. The armed confrontation must reach a *minimum level of intensity*…
>
> (2008: 5, emphasis in original)

Non-international armed conflicts are triggered by a particular set of material factors. Chief among these is the existence of organized armed parties 'engaged in fighting of some intensity' (O'Connell 2015: 67). According to a number of authors, UAV strikes conducted outside the 'hot' battlefield fail to meet the threshold of intensity necessary for an armed conflict to exist. The United States is unable, therefore, to draw upon the more permissive standards of the laws of war to justify its violence (Alston 2010: par 53; Duffy 2015;

O'Connell 2010: par 14). While none of these authors reference radical asymmetry directly, the one-directional nature of UAV-exclusive violence would appear to be of relevance to their determination.

In order to clarify this, consider the 2006 conflict between Israel and Lebanon. According to Lubell 'there was little question' that the intensity of fighting between Israel and Hezbollah met the threshold for an armed conflict (2010: 130). His evidence for this included the extensive casualties endured by Lebanon as a consequence of Israeli aerial bombing and ground operations, as well as the thousands of rockets launched by Hezbollah against Israel (Lubell 2010: 250). In short, there existed a protracted, intense, and *reciprocal* exchange of violence between adversaries. It logically follows that hostilities that lack such reciprocity, including UAV-exclusive violence, will struggle to meet the NIAC threshold of intensity.

There are material constancies to war that if pressured, signify a change not merely in its character, but in its very *nature*. Structural reciprocal risk represents one of these constancies. Its erosion, through the imposition of radically asymmetric conditions of war, challenges our assumptions regarding the moral permissibility of inter-combatant violence.

But is that all it challenges? This book has retained the term 'war' to describe the radically asymmetric violence of the United States. The central reason for this is that while the responsive capacity of those targeted has been severely abridged—more so than at any other time in history—it has not been truly eliminated. The applicability of the term 'war' becomes less certain, however, when we shift our focus to the law, and specifically, NIACs and the legal threshold of intensity.

This section has argued that radical asymmetry necessitates a reformulation of the laws of war, from status to conduct-based targeting. In reaching this determination, however, we should remain open to a further possibility, that the military imbalance under discussion calls into question not merely the coherence, but the very *applicability* of the laws of war. At a minimum, this suggests a further destabilization in the frameworks that legitimize killing in war, as well as a growing inconsistency between what is legal on the battlefield and what is moral. The implications of this destabilization must be grappled with, particularly given the likely trajectory of radically asymmetric violence.

The Trajectory of Radical Asymmetry

A consistent argument throughout this book is that disharmony between the material structure of battle and the assumptions that underpin its moral guidelines has the potential to significantly challenge the way in which we justify killing in war. At the same time, however, we must recognize that this structure is never finalized. The trajectory of the moral challenge of radical asymmetry will be determined largely by how resilient the military imbalances that have created it prove to be.

This imbalance is currently most evident within the context of the UAV-exclusive violence of the United States. This dominance is likely to diminish, however, as more and more states acquire their own arsenal of armed and long-range UAVs. The global proliferation of unmanned technology will be sufficient evidence for some of the *impermanence* of radical asymmetry. The logic being that as other actors obtain comparable military advantage, the United States will be rendered increasingly vulnerable to the geographically boundless, one-directional violence it has hitherto monopolized. Radical military imbalance would conclude in such a scenario, as would, by extension, its tension with the moral justifications for killing in war.

What this overlooks, however, are the actual dynamics of radically asymmetric violence. Risk-asymmetry in war is determined just as much by the military deficiencies of the weaker party as it is by the technological benefits of the strong. The degree of structural imbalance necessary to challenge the moral right to kill on the battlefield is currently, and for the foreseeable future, *only* achievable against non-state actors, who lack the capacity to meaningfully overturn this unfavourable status quo. Going forward, what should concern us is not that states will nullify the unilateral violence of the United States, but rather replicate it, against their own non-state enemies. It is simple enough to imagine future conditions in which military force is directed with impunity by less principled actors, not against groups such as ISIS or al-Qaeda, for whom the term 'criminal' is among the more modest of charges that can be levelled, but rather against individuals and organizations with whom we sympathize. In more extreme cases, this could include political dissidents, socially marginalized groups, and other 'enemies of the state' (Alston 2005: par 41; Barshad 2018).[9]

In grappling with the implications of this emerging violence, this book stands as an important resource. It has first identified the indirect but critical role of reciprocal risk in the moral justifications for lethal force. It has further asserted the existence of a moral crossover point in war, at which the violence of the stronger party may be so one-directional as to undermine the moral

[9] In 2013, China considered the use of a weaponized UAV to kill a suspected drug lord in a remote region of Myanmar (Perlez 2013).

basis for its permissiveness. As other actors obtain the same military capabilities as the United States, we are likely to observe a similar relaxation in the moral criteria for acceptable violence. This book provides a theoretically and historically grounded framework for better determining the point at which such violence is no longer compatible with the assumptions that underpin the moral right to kill.

References

A True Patriot for the Common Good of his Native Country of England. 1974. *A New Invention of Shooting Fire-Shafts in Long-Bowes*. Amsterdam: Theatrum Orbis Terrarum; Norwood, NJ: W. J. Johnson.

ABC News. 2013. 'US Scraps Medal for Drone, Cyber Warriors'. Accessed 25 June 2013. http://www.abc.net.au/news/2013-04-16/us-scraps-medal-for-drone2c-cyber-warriors/4632086.

Abé, Nicola. 2012. 'The Woes of an American Drone Operator'. Accessed 20 May 2017. http://www.spiegel.de/international/world/pain-continues-after-war-for-american-drone-pilot-a-872726.html.

Ackerman, Felicia. 2003. '"Never to Do Outrageousity nor Murder": The World of Malory's Morte Darthur'. In *The Code of the Warrior: Exploring Warrior Values Past and Present*, edited by S. E. French. Oxford: Rowman & Littlefield.

Adams, Michael J. and Ryan Goodman. 2018. '"Reasonable Certainty" vs "Near Certainty" in Military Targeting – What the Law Requires'. *Just Security*. Accessed 10 March 2019. https://www.justsecurity.org/52343/reasonable-certainty-vs-near-certainty-military-targeting-what-law-requires/.

Adas, Michael. 2014. *Machines as the Measure of Men: Science, Technology, and Ideologies of Western Dominance*. Ithaca NY and London: Cornell University Press.

Adorno, Theodor W. 2005. *Minima Moralia: Reflections from Damaged Life*. London: Verso.

Aerospace Magazine. 2013 (November). 'The "Drone" Debate: The Ethics of Armed Unmanned Aircraft'.

Ahmad, Shakeel. 2013. 'A Legal Assessment of the US Drone Strikes in Pakistan'. *International Criminal Law Review* 13 (4): 917–30.

Ahmed, Akbar and Frankie Martin. 2012. 'Deadly Drone Strike on Muslims in the Southern Philippines'. Accessed 11 September 2017. https://www.brookings.edu/opinions/deadly-drone-strike-on-muslims-in-the-southern-philippines/.

Aksoy, Asu and Kevin Robins. 1991. 'Exterminating Angels: Morality, Violence and Technology in the Gulf War'. *Science as Culture* 2 (3): 322–36.

Alexander, David. 2013a. 'U.S. Navy Makes Aviation History with Carrier Drone Launch'. Accessed 10 March 2014. http://www.reuters.com/article/2013/05/14/usa-navy-drone-idUSL2N0DV2VA20130514.

Alexander, David. 2013b. 'Retired General Cautions Against Overuse of "Hated" Drones'. Accessed 13 May 2017. http://www.reuters.com/article/us-usa-afghanistan-mcchrystal-idUSBRE90608O20130107.

Alexander, Lawrence A. 1976. 'Self-Defense and the Killing of Noncombatants: A Reply to Fullinwider'. *Philosophy & Public Affairs* 5 (4): 408–15.

Allen, H. R. 1972. *The Legacy of Lord Trenchard*. London: Cassell.

Allen, Thomas B., F. Clifton Berry, and Norman Polmar. 1991. *War in the Gulf*. Atlanta, GA: Turner Publishing.

Alston, Philip. 2005. 'Report of the Special Rapporteur on Extrajudicial, Summary, or Arbitrary Execution'. Accessed 12 July 2017. https://documents-dds-ny.un.org/doc/UNDOC/GEN/G05/101/34/PDF/G0510134.pdf?OpenElement.

Alston, Philip. 2010. 'Report of the Special Rapporteur on Extrajudicial, Summary or Arbitrary Executions'. Accessed 18 September 2015. http://www2.ohchr.org/english/bodies/hrcouncil/docs/14session/A.HRC.14.24.Add6.pdf.

Amnesty International. 2000. 'NATO/Federal Republic of Yugoslavia "Collateral Damage" or Unlawful Killings? Violations of the Laws of War by NATO during Operation Allied Force'. Accessed 20 July 2013. https://www.amnesty.org/download/Documents/140000/eur700182000en.pdf.

Anderson, Kenneth. 2009. 'Targeted Killing in U.S. Counterterrorism Strategy and Law'. Accessed 22 August 2013. https://ssrn.com/abstract=1415070.

Anderson, Kenneth. 2012. 'Efficiency in Bello and ad Bellum: Making the Use of Force Too Easy?' In *Targeted Killings: Law and Morality in an Asymmetric World*, edited by J. D. Ohlin, A. Altman, and C. Finkelstein. Oxford: Oxford University Press.

Anderson, Kenneth. 2014. 'Laurie Blank on Proportionality in the International Law of Targeting'. Accessed 3 November 2016. https://www.washingtonpost.com/news/volokh-conspiracy/wp/2014/07/31/laurie-blank-on-proportionality-in-the-international-law-of-targeting/.

Anderson, R. Wherry. 1917. *The Romance of Air-Fighting*. London: Cassell and Co.

Anscombe, G. E. M. 1981a. 'Mr. Truman's Degree'. In *The Collected Philosophical Papers of G. E. M. Anscombe*. Vol. 3. Oxford: Basil Blackwell.

Anscombe, G. E. M. 1981b. *The Collected Philosophical Papers of G. E. M. Anscombe*. Vol. 3. Oxford: Basil Blackwell.

Anscombe, G. E. M. 2006. 'War and Murder'. In *The Ethics of War: Classical and Contemporary Readings*, edited by H. Syse, E. Begby, and G. M. Reichberg. Malden, MA: Blackwell Publishing.

Applbaum, Arthur. 1999. *Ethics of Adversaries: The Morality of Roles in Public and Professional Life*. Princeton, NJ: Princeton University Press.

Arkin, Ronald. 2010. 'The Case for Ethical Autonomy in Unmanned Systems'. *Journal of Military Ethics* 9 (4): 332–41.

Asad, Talal. 2007. *On Suicide Bombing*. New York: Columbia University Press.

Asaro, Peter. 2008. 'How Just Could a Robot War Be?'. In *Current Issues in Computing and Philosophy*, edited by A. Briggle, K. Waelbers, and P. Brey. Amsterdam: IOS Press.

Augustine, Saint. 1950. *The City of God*. New York: Modern Library.

Augustine, Saint. 2009. 'Reply to Faustus the Manichaean: Book XXII, para 75'. Accessed 20 June 2014. http://www.gnosis.org/library/contf2.htm.

Axe, David. 2012. 'How to Prevent Drone Pilot PTSD: Blame the "Bot"'. Accessed 11 November 2015. http://www.wired.com/2012/06/drone-pilot-ptsd/.

Ayala, Balthazar. 1912. *De Jure et Officiis Bellicis et Disciplina Militari Libri III*. Edited by J. Westlake. Translated by J. P. Bate. Washington, DC: Carnegie Institution of Washington.

Bacevich, Andrew J. 2005. *The New American Militarism: How Americans Are Seduced by War*. Oxford: Oxford University Press.

Baker, Newton. 1919. *Annual Report of the Secretary of War*. Jones Aviation Chronology, July–December 1919.

Balabanova, Ekaterina. 2007. *Media, Wars, and Politics: Comparing the Incomparable in Western and Eastern Europe*. Aldershot: Ashgate Publishing.

Balderston, Marion and David Syrett. 1975. *The Lost War: Letters from British Officers During the American Revolution*. New York: Horizon Press.

Baldwin, Hanson W. 1946. 'The "Drone": Portent of Push-Button War: Recent Operations Point to the Pilotless Plane as a Formidable Weapon in War's New Armory'. *The New York Times*, 25 August.

Barela, Steven J. and Avery Plaw. 2016. 'Are Drones Less Accurate than Piloted Aircraft?' Accessed 11 May 2017. https://www.justsecurity.org/33333/drones-accurate-piloted-aircraft-2/.

Barrett, Frank. 1996. 'The Organizational Construction of Hegemonic Masculinity: The Case of the US Navy'. *Gender, Work & Organisation* 3 (3): 129–42.

Barrinha, Andre and Luis da Vinha. 2015. 'Dealing with Risk: Precision Strikes and Interventionism in the Obama Administration'. In *Precision Strike Warfare and International Intervention: Strategic, Ethico-Legal, and Decisional Implications*, edited by W. Aslam, M. Aaronson, T. Dyson, and R. Rauxloh. Abingdon: Routledge.

Barshad, Amos. 2018. 'Extraordinary Measures: Israel's Gabriella Blum Helped Write the Laws of Drone Warfare. Nearly Two Decades Later, She Has Regrets'. Accessed 10 January 2019. https://theintercept.com/2018/10/07/israel-palestine-us-drone-strikes/.

Barwick, Humfrey. 1592. *A Breefe Discourse, Concerning the Force and Effect of all Manuall Weapons of Fire, and the Disability of the Long Bowe or Archery, in Respect of Others of Greater Force Now in Vse*. Accessed 13 July 2019. https://quod.lib.umich.edu/e/eebo/A05277.0001.001?rgn=main;view=fulltext.

Baudrillard, Jean. 1991. *The Gulf War Did Not Take Place*. Translated by P. Patton. Bloomington, IN: Indiana University Press.

Bauman, Zygmunt. 1989. *Modernity and the Holocaust*. Cambridge: Polity Press.

BBC News. 1999. 'World: Europe Analysis: Sending in the Troops'. Accessed 17 May 2013. http://news.bbc.co.uk/1/hi/world/europe/317696.stm.

BBC News. 2013. 'US Drone Programme: "Strict, Fair and Accountable" – Kerry'. Accessed 15 May 2017. http://www.bbc.com/news/av/world-radio-and-tv-22690918/us-drone-programme-strict-fair-and-accountable-kerry.

BBC News. 2015. 'Islamic State Conflict: Two Britons Killed in RAF Syria Strike'. Accessed 10 November. http://www.bbc.co.uk/news/uk-34178998.

BBC News. 2017. 'Islamic State: UK Military Could Honour Drone Pilots'. Accessed 21 September 2017. http://www.bbc.com/news/uk-41329782.

Bearak, Barry. 2001. 'A Nation Challenged: Death on the Ground; U.S. Raid Kills Unknown Number in an Afghan Village'. Accessed 15 October 2015. http://www.nytimes.com/2001/10/13/world/nation-challenged-death-ground-us-raid-kills-unknown-number-afghan-village.html.

Beauchamp, Zack and Julian Savulescu. 2013. 'Robot Guardians: Teleoperated Combat Vehicles in Humanitarian Military Intervention'. In *Killing by Remote Control: The Ethics of an Unmanned Military*, edited by B. J. Strawser. Oxford: Oxford University Press.

Beck, Ulrich. 1999. *World Risk Society*. Cambridge: Polity Press.

Becker, Joe and Scott Shane. 2012. 'Secret "Kill List" Proves a Test of Obama's Principles and Will'. *The New York Times*, 29 May, A1.

Beevor, Antony. 1999. 'The Soldiers Preferred Margaret'. Accessed 13 April 2015. https://www.newstatesman.com/node/192882.

Bell, David A. 2012. 'In Defense of Drones: A Historical Argument'. Accessed 12 October 2015. https://newrepublic.com/article/100113/obama-military-foreign-policy-technology-drones.

Bell, Duncan S. A. 2003. 'Political Theory and the Functions of Intellectual History: A Response to Emmanuel Navon'. *Review of International Studies* 29 (1): 151–60.

Bellamy, Alex J. 2006. *Just Wars: From Cicero to Iraq*. Cambridge: Polity Press.

Benbaji, Yitzhak. 2008. 'A Defense of the Traditional War Convention'. *Ethics* 118 (3): 464–95.

Benbaji, Yitzhak. 2011. 'The Moral Power of Soldiers to Undertake the Duty of Obedience'. *Ethics* 122 (1): 43–73.

Bergen, Peter and Jennifer Rowland. 2012. 'CIA Drone War in Pakistan in Sharp Decline'. Accessed 11 September 2017. https://edition.cnn.com/2012/03/27/opinion/bergen-drone-decline/index.html.

Bergen, Peter and Katherine Tiedemann. 2010. 'The Year of the Drone: An Analysis of U.S. Drone Strikes in Pakistan, 2004–2010'. New America Foundation.

Bertaud, Jean P. 1988. *The Army of the Revolution: From Citizen-Soldiers to Instruments of Power*. Princeton, NJ: Princeton University Press.

Best, Geoffrey. 1980. *Humanity in Warfare: The Modern History of the International Law of Armed Conflicts*. London: Weidenfeld & Nicolson.

Best, Geoffrey. 1986. *War and Society in Revolutionary Europe, 1770–1870*. New York: Oxford University Press.

Bialer, Uri. 1980. *The Shadow of the Bomber: The Fear of Air Attack and British Politics*. London: Royal Historical Society.

Biddle, Tami Davis. 1994. 'Air Power'. In *The Laws of War: Constraints on Warfare in the Western World*, edited by G. J. Andreopoulos, M. R. Shulman, and M. Howard. New Haven, CT: Yale University Press.

Biddle, Tami Davis. 1995. 'British and American Approaches to Strategic Bombing: Their Origins and Implementation in the World War II Combined Bomber Offensive'. *Journal of Strategic Studies* 18 (1): 91–144.

Biddle, Tami Davis. 2008. 'Dresden 1945: Reality, History, and Memory'. *The Journal of Military History* 72 (2): 413–49.

Biddle, Tami Davis. 2014. 'Strategic Bombardment: Expectation, Theory, and Practice in the Early Twentieth Century'. In *The American Way of Bombing: Changing Ethical and Legal Norms, from Flying Fortresses to Drones*, edited by M. Evangelista and H. Shue. Ithaca, NY: Cornell University Press.

Bijnkershoek, Cornelius van. 1810. *A Treatise on the Law of War*. Philadelphia: Fry and Kammerer.

Bin Laden, Osama. 2003. 'Bin Laden Audiotape, Sermon on the First Day of Id al-Adha, 16 February 2003'. In *Al-Qaida Statements 2003–2004: A Compilation of Translated Texts by Usama bin Ladin and Ayman al-Zawahiri*, edited by T. Hegghammer. Kjeller: Norwegian Defence Research Establishment.

Blackburn, John. 1917. 'Science of Sniping Now Part of Kearny's War Curriculum'. *Los Angeles Times*, 9 December, 28.

Blair, Dave. 2012. 'Ten Thousand Feet and Ten Thousand Miles: Reconciling our Air Force Culture to Remotely Piloted Aircraft and the New Nature of Aerial Combat'. *Air and Space Power* 26 (3): 61–9.

Blair, Dennis. 2011. 'Drones Alone Are Not the Answer'. *The New York Times*, 15 August, A1.

Blank, Laurie R. 2015. 'Targeted Killing'. In *The Ashgate Research Companion to Military Ethics*, edited by J. T. Johnson and E. D. Patterson. Farnham: Ashgate.

Blount, Clive. 2011. 'War at a Distance? Some Thoughts for Airpower Practitioners'. *Air Power Review* 14 (2): 31–41.

Blum, Gabriella. 2010. 'The Dispensable Lives of Soldiers'. *Journal of Legal Analysis* 2 (1): 69–124.

Bollet, Alfred J. 1992. 'Scurvy and Chronic Diarrhea in Civil War Troops: Were They Both Nutritional Deficiency Syndromes?' *Journal of the History of Medicine and Allied Sciences* 47 (1): 49–67.

Bone, Elizabeth and Christopher Bolkom. 2003. 'Unmanned Aerial Vehicles: Background and Issues for Congress'. Accessed 11 September 2016. https://fas.org/irp/crs/RL31872.pdf.

Boot, Max. 2002. *The Savage Wars of Peace: Small Wars and the Rise of American Power*. New York: Basic Books.

Booth, Ken. 2000. 'The Kosovo Tragedy: Epilogue to Another "Low and Dishonest Decade"'. *Politikon* 27 (1): 5–18.

Boothby, William H. 2009. *Weapons and the Law of Armed Conflict*. Oxford: Oxford University Press.

Boothby, William H. 2014. 'The Law of Targeting'. In *Contemporary Challenges to the Laws of War: Essays in Honour of Professor Peter Rowe*, edited by J. Summers, N. D. White, and C. Harvey. Cambridge: Cambridge University Press.

Bosco, David. 2008. 'Moral Principle vs. Military Necessity'. *The American Scholar* 77 (1): 25–34.

Bourke, Joanna. 1999. *An Intimate History of Killing: Face-to-Face Killing in the Twentieth Century*. London: Granta Books.

Bowcott, Owen. 2012. 'Drone Attacks in Pakistan are Counterproductive, Says Report'. *The Guardian*, 25 September, 19.

Boyle, Alan and C. M. Chinkin. 2007. *The Making of International Law*. Oxford: Oxford University Press.

Boyle, M. J. 2015. 'The Legal and Ethical Implications of Drone Warfare'. *The International Journal of Human Rights* 19 (2): 105–26.

Bradley, J. P. 2010. 'An Exploratory Study on Sniper Well-Being: Report on the First Year of Sniper Well-Being Research (FY 2008-2009)'. Accessed 12 September 2017. http://cradpdf.drdc-rddc.gc.ca/PDFS/unc102/p534015_A1b.pdf.

Brandt, Richard B. 1972. 'Utilitarianism and the Rules of War'. *Philosophy & Public Affairs* 1 (2): 145–65.

Braun, Megan and Daniel Brunstetter. 2013. 'Rethinking the Criterion for Assessing CIA-Targeted Killings: Drones, Proportionality and Jus Ad Vim'. *Journal of Military Ethics* 12 (4): 304–24.

Bray, Hiawatha. 2007. 'Debuting in Boston: A 250-Pound Robot to Help Soldiers Stay Alive'. *Boston Globe*, 14 May.

Brennan, John. 2012. 'The Ethics and Efficacy of the President's Counterterrorism Strategy'. Accessed 11 August 2017. https://www.wilsoncenter.org/event/the-efficacy-and-ethics-us-counterterrorism-strategy.

Brew, Terry. 2004. 'Lucky Me: Part Two: Defence of the South Coast, D-Day, Post-D-Day Fighting'. Accessed 11 May 2017. http://www.bbc.co.uk/history/ww2peopleswar/stories/83/a2228483.shtml.

Brooks, Rosa. 2012a. 'What's Not Wrong With Drones? The Wildly Overblown Case Against Remote-Controlled War'. Accessed 13 August 2014. http://www.foreignpolicy.com/articles/2012/09/05/whats_not_wrong_with_drones.

Brooks, Rosa. 2012b. 'Take Two Drones and Call Me in the Morning'. Accessed 13 January 2013. http://www.foreignpolicy.com/articles/2012/09/12/take_two_drones_and_call_me_in_the_morning.

Brooks, Rosa. 2013. 'Drones and Cognitive Dissonance'. Accessed 24 March 2016. https://scholarship.law.georgetown.edu/cgi/viewcontent.cgi?article=2266&context=facpub.

Brooks, Rosa. 2014. 'Duck-Rabbits and Drones: Legal Indeterminacy in the War on Terror'. *Stanford Law and Policy Review* 25 (2): 301–16.

Brubaker, Rogers. 1984. *The Limits of Rationality: An Assay on the Social and Moral Thought of Max Weber*. London: Allen & Unwin.

Brunstetter, Daniel. 2012. 'Can We Wage a Just Drone War?' Accessed 15 January 2013. http://www.theatlantic.com/technology/archive/2012/07/can-we-wage-a-just-drone-war/260055/.

Buley, Benjamin. 2008. *The New American Way of War*. London and New York: Routledge.

Bull, Hedley. 1966. 'International Theory: The Case for a Classical Approach'. *World Politics* 18 (3): 361–77.

Bull, Hedley. 1977. *The Anarchical Society: A Study of Order in World Politics*. London: Macmillan.

Burbidge, William F. 1946. *From Balloon to Bomber: A Complete History of Aviation from Earliest Times Until the Present Day*. London: J. Crowther.

Burrell, Joseph W. 2011. *Work is for the People: A Treatise on the -Isms*. New York: Algora Publishing.

Busch, Briton Cooper. 2006. *Bunker Hill to Bastogne: Elite Forces and American Society*. Washington, DC: Potomac Books.

Bush, George W. 2003. 'President Rallies Troops at Fort Hood'. Accessed 13 May 2017. https://georgewbush-whitehouse.archives.gov/news/releases/2003/01/20030103.html.

Byers, S. H. M. 1896. 'How Men Feel in Battle: Recollections of a Private at Champion Hills'. *The Annals of Iowa* 2 (6): 438–49.

Canning, John. 2008. 'Weaponized Unmanned Systems: A Transformational Warfighting Opportunity, Government Roles in Making it Happen'. *Proceedings of Engineering the Total Ship (ETS)*: 23–5.

Cardot, Lucien. 1908. *Hérésies et Apostasies Militaires de Notre Temps*. Paris: Berger-Levrault.

Carnahan, Burrus M. 2012. 'The Civil War Origins of the Modern Rules of War'. *Northern Kentucky Law Review* 39 (4): 661–97.

Cartledge, Paul. 2002. *The Greeks: A Portrait of Self and Others*. Oxford: Oxford University Press.

Carus, Seth W. 1982. 'Bekaa Valley Campaign'. *The Washington Quarterly* 5 (4): 34–41.

Casey-Maslen, Stuart. 2014. 'The Use of Armed Drones'. In *Weapons under International Human Rights Law*, edited by S. Casey-Maslen. Cambridge: Cambridge University Press.

Center for a New American Security. 2016. 'U.S. Public Support for Drone Strikes: When do Americans Prefer Unmanned over Manned Platforms?' Accessed 11 January 2019. https://www.cnas.org/publications/reports/u-s-public-support-for-drone-strikes.

Central Intelligence Agency. 2001. 'Intelligence Community Assessment of the Lieutenant Commander Speicher Case'. Accessed 12 August 2017. https://web.archive.org/web/20041017122850/http:/www.foia.cia.gov/docs/DOC_0000588922/0000588922_0001.gif.

Chait, Jonathan. 2016. 'Five Days that Shaped a Presidency'. Accessed 12 May 2017. http://nymag.com/daily/intelligencer/2016/10/barack-obama-on-5-days-that-shaped-his-presidency.html.

Chamayou, Grégoire. 2012. *Manhunts: A Philosophical History*. Princeton, NJ: Princeton University Press.

Chamayou, Grégoire. 2015. *Drone Theory*. London: Penguin Books.

Chant, Christopher. 1996. *Warfare and the Third Reich: The Rise and Fall of Hitler's Forces*. London: Salamander.

Charnwood, G. R. B. 1996. *Abraham Lincoln*. London: Madison Books.

Chase, Kenneth. 2008. *Firearms: A Global History to 1700*. New York and Cambridge: Cambridge University Press.

Chediac, Joyce. 1991. 'The Massacre of Withdrawing Soldiers on "The Highway of Death"'. Accessed 16 April 2017. http://www.jacobsm.com/deoxy/deoxy.org/wc/wc-death.htm.

Chenoweth, Erica. 2012. 'Do Drones Change American's Views on the Use of Force'. Accessed 15 January 2013. http://themonkeycage.org/blog/2012/08/28/do-drones-change-americans-views-on-the-use-of-force/.

Chesney, Robert. 2016. 'Shift to JSOC on Drone Strikes Does Not Mean CIA Has Been Sidelined'. Accessed 11 September 2017. https://www.lawfareblog.com/shift-jsoc-drone-strikes-does-not-mean-cia-has-been-sidelined.

Childress, James F. 1976. 'Francis Lieber's Interpretation of the Laws of War: General Order No. 100 in the Context of His Life and Thought'. *American Journal of Jurisprudence* 21 (1): 34–70.

Christensen, Thomas J. and Jack Snyder. 1990. 'Chain Gangs and Passed Bucks: Predicting Alliance Patterns in Multipolarity'. *International Organization* 44 (2): 137–68.

Christopher, John. 1956. *Balloons at War: Gasbags, Flying Bombs and Cold War Secrets*. Stroud: Tempus.

Christopher, Paul. 2004. *The Ethics of War and Peace: An Introduction to Legal and Moral Issues*. Upper Saddle River, NJ: Prentice Hall.

Chun, Lisa. 2006. 'Robotic Technology Lowers Military Risk'. Accessed 20 June 2013. http://www.upi.com/inc/view.php?StoryID=20060605-053418-6447r.

Churchill, Winston S. 1902. *The River War: An Historical Account of the Reconquest of the Soudan*. London: Longmans, Green, and Co.

Churchill, Winston S. 2007. *My Early Life: A Roving Commission*. London: The Folio Society.

Cicero, Marcus Tullius. 1850. *Cicero's Three Books of Offices, or Moral Duties*. Translated by C. R. Edmonds. London: H. G. Bohn.

Clark, Ian. 1988. *Waging War: A Philosophical Introduction*. Oxford: Clarendon Press.

Clark, Ian. 2013. *The Vulnerable in International Society*. Oxford: Oxford University Press.

Clark, Ian. 2015. *Waging War: A New Philosophical Introduction*. Vol. 2. Oxford: Oxford University Press.

Clark, John W. 1964. 'Remote Control in Hostile Environments'. *New Scientist* 22 (389): 300–3.

Clausewitz, Carl von. 2008. *On War*. Translated by M. Howard and P. Parat. Oxford: Oxford University Press.

Clinton, Bill. 1999. 'Statement on Kosovo (March 24, 1999)'. Accessed 11 May 2013. https://millercenter.org/the-presidency/presidential-speeches/march-24-1999-statement-kosovo.

Cloud, David S. 2010. 'CIA Drones Have Broader List of Targets'. Accessed 11 November 2015. http://articles.latimes.com/2010/may/05/world/la-fg-drone-targets-20100506.

CNN. 2009. 'Remote Warfare Ushers New Kind of Stress'. Accessed 10 August 2017. http://edition.cnn.com/2009/WORLD/americas/07/23/wus.warfare.pilots.uav/index.html.

Coady, C. A. J. 2008a. 'The Status of Combatants'. In *Just and Unjust Warriors*, edited by D. Rodin and H. Shue. Oxford: Oxford University Press.

Coady, C. A. J. 2008b. *Morality and Political Violence*. New York and Cambridge: Cambridge University Press.

Coates, Anthony. 1997. *The Ethics of War*. Manchester: Manchester University Press.

Coates, Anthony. 2008. 'Is the Independence Application of Jus in Bello the Way to Limit War?' In *Just and Unjust Warriors: The Moral and Legal Status of Soldiers*, edited by D. Rodin and H. Shue. Oxford: Oxford University Press.

Cockburn, Alexander. 1991. 'Column Left: Unlimited Violence Wins Out: In This Or any U.S. War, It's Delusional to think that Civilians Might Be Spared'. Accessed 11 March 2015. http://articles.latimes.com/1991-03-11/local/me-15_1_korean-war.

Cockburn, Alexander. 2015. *Kill Chain: Drones and the Rise of High-Tech Assassins*. London: Verso.

Coe, Robert and Michael Schmitt. 1997. 'Fighter Ops for Shoe Clerks'. *The Air Force Review* 42: 49–99.

Cohen, Eliot. 1994. 'The Mystique of U.S. Air Power'. *Foreign Affairs* 73 (1): 109–24.

Cohen, Richard. 2009. 'Is the Afghanistan Surge Worth the Lives That Will Be Lost'. *Washington Post*, 8 December.

Cohen, Sheldon. 1989. *Arms and Judgment: Law, Morality, and the Conduct of War in the Twentieth Century*. Boulder, CO: Westview.

Cohen, William S. and Gen. Henry H. Shelton. 1999. *Joint Statement on the Kosovo After Action Review*. Washington, DC: Office of Assistant Secretary of Defense.

Coker, Christopher. 2001a. *Humane Warfare: The New Ethics of Postmodern War*. London and New York: Routledge.

Coker, Christopher. 2001b. 'The United States and the Ethics of Post-Modern'. In *Ethics and Foreign Policy*, edited by K. E. Smith and M. Light. New York and Cambridge: Cambridge University Press.

Coker, Christopher. 2002a. 'Globalisation and Insecurity in the Twenty-First Century: NATO and the Management of Risk'. *Adelphi Paper* 42 (345): 57–64.

Coker, Christopher. 2002b. *Waging War Without Warriors? The Changing Culture of Military Conflict*. Boulder, CO: Lynne Rienner.

Coker, Christopher. 2007. *The Warrior Ethos: Military Culture and the War on Terror* Abingdon: Routledge.

Coker, Christopher. 2010. *Barbarous Philosophers: Reflections on the Nature of War from Heraclitus to Heisenberg*. London: Hurst.

Coker, Christopher. 2014. *Can War be Eliminated?* Cambridge: Polity Press.

Colby, Elbridge. 1927. 'How to Fight Savage Tribes'. *American Journal of International Law* 21: 279–88.

Cole, Chris. 2017. 'Interview of Former RAF Reaper Pilot "Justin Thompson" (a pseudonym) by Chris Cole, Drone Wars UK, May 2017'. Accessed 13 August 2017. https://dronewarsuk.files.wordpress.com/2017/05/justin-thompson-interview-transcript.pdf.

Cole, Christopher and E. F. Cheesman. 1984. *The Air Defence of Britain*. London: Bodley Head.

Cole, C., M. Dobbing, and A. Hailwood. 2010. *Convenient Killing: Armed Drones and the 'PlayStation' Mentality*. Oxford: Fellowship of Reconciliation.

Cole, David. 2016. 'The Drone Presidency'. Accessed 19 January 2017. http://www.nybooks.com/articles/2016/08/18/the-drone-presidency/.

Condren, Conal. 2014. 'Historiographical Myth, Discipline, and Contextual Distortion'. *History of European Ideas* 40 (1): 37–43.

Connell, R. W. 1985. 'Masculinity, Violence, and War'. In *War/Masculinity*, edited by P. Patton and R. Poole. Sydney: Intervention.

Constant, Benjamin. 1988. 'Spirit of Conquest'. In *Political Writings*, edited by B. Fontana. Cambridge: Cambridge University Press.

Cook, Martin. 2004. *The Moral Warrior: Ethics and Service in the US Military*. Albany, NY: State University of New York Press.

Coppard, George. 1969. *With a Machine Gun to Cambrai: The Tale of a Young Tommy in Kitchener's Army 1914–1918*. London: HMSO.

Cordesman, Anthony H. 2001. *The Lessons and Non-Lessons of the Air and Missile Campaign in Kosovo*. Westport, CT: Praeger.

Corn, Joseph J. 1983. *The Winged Gospel: America's Romance with Aviation, 1900–1950*. New York: Oxford University Press.

Cortright, David and Rachel Fairhurst. 2015. 'Assessing the Debate on Drone Warfare'. In *Drones and the Future of Conflict: Ethical, Legal, and Strategic Implications*, edited by R. Fairhurst, K. Wall, and D. Cortright. Chicago, IL: University of Chicago Press.

Costello, Edward. 1841. *The Adventures of a Soldier: Or, Memoirs of Edward Costello K.S.F.* London: Henry Colburn Publisher.

Cousins, Norman. 1978. 'Editor's Odyssey'. *Saturday Review*, 15 April.

Croker, John W. 1885. *The Croker Papers: The Correspondence and Diaries of the Late Right Honourable John Wilson Croker, Secretary to the Admiralty from 1809 to 1830*. Edited by L. J. Jennings. London: John Murray.

Crootof, Rebecca. 2019. 'Regulating New Weapons Technology'. In *The Impact of Emerging Technologies on the Law of Armed Conflict*, edited by Eric Talbot Jensen and Ronald T. P. Alcala. New York: Oxford University Press.

Crozier, F. P. 1937. *The Men I Killed*. London: Michael Joseph.

Crozier, F. P. 1950. *A Brass Hat in No Man's Land*. London: C. Chivers.

Currier, Cora and Justin Elliot. 2013. 'The Drone Doctrine We Still Know Nothing About'. Accessed 9 May 2017. https://www.propublica.org/article/drone-war-doctrine-we-know-nothing-about.

Darwin, Charles. 1889. *The Descent of Man and Selection in Relation to Sex*. New York: Appleton and Co.

Daskal, Jennifer. 2013. 'The Geography of the Battlefield: A Framework for Detention and Targeting Outside the "Hot" Conflict Zone'. *University of Pennsylvania Law Review* 161 (5): 1165–34.

Dawson, Doyne. 1996. *The Origins of Western Warfare: Militarism and Morality in the Ancient World*. Boulder, CO: Westview Press.

Deane, Herbert Andrew. 1963. *The Political and Social Ideas of Saint Augustine*. New York: Columbia University Press.

Debrabander, Firmin. 2014. 'Drones and the Democracy Disconnect'. *The New York Times*, 14 September.

Deforest, William. 1867. 'A Union Officer's Personal Account of the Siege'. Accessed 12 April 2015. http://pth.thehardyparty.com/deforest.htm.

DeGroot, Gerard J. 1999. 'A Noble Cause? America and the Vietnam War'. In *Major Problems in the History of the Vietnam War*, edited by R. McMahon. Boston, MA: Houghton Mifflin.

Delahunty, Robert J. and John C. Yoo. 2002. 'The President's Constitutional Authority to Conduct Military Operations Against Terrorist Organizations and the Nations that Harbour or Support Them'. *Harvard Journal of Law and Public Policy* 25 (2): 487–517.

Demetriou, Daniel. 2013. 'Honor War Theory: Romance or Reality?' *Philosophical Papers* 42 (3): 285–313.

Dempsey, Martin E. 2010. 'An Interview with Martin E. Dempsey'. *Prism* 2 (1): 151–5.

Der Derian, James. 2009. *Virtuous War: Mapping the Military-Industrial-Media-Entertainment Network*. New York and London: Routledge.

Devereaux, Ryan. 2015. 'Manhunting in the Hindu Kush: Civilian Casualties and Strategic Failures in American's Longest War'. Accessed 10 November 2016. https://theintercept.com/drone-papers/manhunting-in-the-hindu-kush/.

Devereaux, Ryan and Alex Emmons. 2016. 'Obama Worries Future Presidents will Wage Perpetual, Covert Drone War'. Accessed 16 October 2016. https://theintercept.com/2016/10/03/obama-worries-future-presidents-will-wage-perpetual-covert-drone-war/.

Devetak, Richard. 2008. 'Globalization's Shadow'. In *The Globalization of Political Violence: Globalization's Shadow*, edited by R. Devetak and C. W. Hughes. London and New York: Routledge.

DeYoung, Karen. 2011. 'CIA Idles Drone Flights from Base in Pakistan'. Accessed 12 September 2016. https://www.washingtonpost.com/world/national-security/cia-idles-drone-flights-from-base-in-pakistan/2011/07/01/AGpOiKuH_story.html.

Dige, Morten. 2012. 'Explaining the Principle of *Mala in Se*'. *Journal of Military Ethics* 11 (4): 318–32.

Dill, Janina. 2015a. 'The 21st-Century Belligerent's Trilemma'. *The European Journal of International Law* 26 (1): 83–108.

Dill, Janina. 2015b. *Legitimate Targets? Social Construction, International Law and US Bombing*. Cambridge: Cambridge University Press.

Dill, Janina and Henry Shue. 2012. 'Limiting the Killing in War: Military Necessity and the St. Petersburg Assumption'. *Ethics & International Affairs* 26 (3): 311–33.

Dinstein, Yoram. 2002. 'Legal and Ethical Lessons of NATO's Kosovo Campaign'. In *Legal and Ethical Lessons of NATO's Kosovo Campaign*, edited by A. E. Wall. Newport, RI: Naval War College.

Dinstein, Yoram. 2007. 'The System of Groups in International Humanitarian Law'. In *International Humanitarian Law Facing New Challenges: Symposium in Honour of Knut Ipsen*, edited by W. H. von Heinegg and V. Epping. Berlin: Springer.

Dinstein, Yoram. 2011. 'LOAC and Attempts to Abuse or Subvert it'. In *International Law and the Changing Character of War*, edited by R. A. Pedrozo and D. P. Wollschlaeger. Newport, RI: Naval War College.

Divine, Robert A. 2000. 'The Persian Gulf War Revisited: Tactical Victory, Strategic Failure'. *Diplomatic History* 24 (1): 129–38.

Dougherty, Kimberly K. 2013. '"Of the Sky Above You Must Beware": Airspace and Airpower in Twentieth Century Literature'. PhD thesis, University of New Hampshire (3572930).

Douhet, Giulio. 2009. *The Command of the Air*. Translated by D. Ferrari. Tuscaloosa, AL: University of Alabama Press.

Dower, John W. 1987. *War Without Mercy: Race and Power in the Pacific War*. New York: Pantheon.

Doyle, Michael. 1983. 'Kant, Liberal Legacies and Foreign Affairs'. *Philosophy & Public Affairs* 12 (3): 205–35.

Draper, G. I. A. D. 1998. *Reflections on Law and Armed Conflicts: The Selected Works on the Laws of War*. The Hague and London: Kluwer Law International.

Duffy, Helen. 2015. *The War on Terror and the Framework of International Law*. Vol. 2. Cambridge: Cambridge University Press.

Duggan, Aaron. 2010. 'Nex Ex Machina, or Bringing it All Back Home: The Fallacy and Fantasy of Sacrificeless Warfare and Why it Will Never Work'. *Mythological Studies Journal* 1 (1). http://journals.sfu.ca/pgi/index.php/pacificamyth/article/view/2/37.

Dunlap, Charles J. 2001. 'The Revolution in Military Legal Affairs: Air Force Legal Professionals in 21st Century Conflicts'. *The Air Force Review* 51: 293–309.

Dunlap, Charles J. 2007. 'Air-Minded Considerations for Joint Counterinsurgency Doctrine'. *Air and Space Power* 21 (4): 63–74.

Dunlap, Charles J. 2014. 'Clever or Clueless? Observations about Bombing Norm Debates'. In *The American Way of Bombing: Changing Ethical and Legal Norms, from Flying Fortresses to Drones*, edited by M. Evangelista and H. Shue. Ithaca, NY: Cornell University Press.

Dunn, D. H. 2013. 'Drones: Disembodied Aerial Warfare and the Unarticulated Threat'. *International Affairs* 89 (5): 1237–46.

Dyer, Gwynne. 1999. 'Biggest Cowards in the World'. *Montreal Gazette*, 11 May, B3.

Earley, Gerald L. 2014. *The Second United States Sharpshooters in the Civil War: A History and Roster*. Jefferson, NC: McFarland.

Echevarria, Antulio. 2007. *Imagining Future War: The West's Technological Revolution and Visions of Wars to Come, 1880–1914*. Westport, CT: Praeger.

Edmunds, Sterling E. 1916. 'The Passing of Chivalry in Warfare: Survivals of the Knightly Code Are Vanishing as Fighting Becomes More Impersonal – The Turk Improves His Reputation'. Accessed 10 April 2015. https://www.nytimes.com/1916/03/29/archives/the-passing-of-chivalry-in-warfare-survivals-of-the-knightly-code.html?searchResultPosition=1

Elias, Norbert. 1994. *The Civilizing Process: The History of Manners and State Formation and Civilization*. Oxford: Blackwell.

Ellis, John. 1986. *The Social History of the Machine Gun*. Baltimore, MD: Johns Hopkins University Press.

Ellul, Jacques. 1965. *The Technological Society*. Translated by J. Wilkinson. London: Jonathan Cape.

Elman, Colin and Miriam Fendius Elman. 2001. 'Introduction: Negotiating International History and Politics'. In *Bridges and Boundaries: Historians, Political Scientists, and the Study of International Relations*, edited by C. Elman and M. F. Elman. Cambridge, MA: MIT Press.

Emprey, Arthur G. 1918. *First Call: Guide Posts to Berlin*. New York and London: G. P. Putnam's Sons.

Enemark, Christian. 2013a. *Armed Drones and the Ethics of War: Military Virtue in a Post-Heroic Age*. London: Routledge.

Enemark, Christian. 2013b. 'The End of Courage? How Drones are Undermining Military Virtue'. Accessed 8 February 2016. http://www.abc.net.au/religion/articles/2013/04/24/3744693.htm.

Entous, Adam. 2010. 'Special Report: How the White House Learned to Love the Drone'. Accessed 20 May 2017. http://www.reuters.com/article/us-pakistan-drones-idUSTRE64H5SL20100518.

Entous, Adam, Siobhan Gorman, and Julian E. Barnes. 2012. 'U.S. Relaxes Drone Rules: Obama Gives CIA, Military Greater Leeway in Use Against Militants in Yemen'. Accessed 10 May 2017. https://www.wsj.com/articles/SB10001424052702304723304577366251852418174.

Eppstein, John. 1935. *The Catholic Tradition of the Law of Nations*. Washington, DC: Catholic Association for International Peace.

Erskine, Toni. 2008. *Embedded Cosmopolitanism: Duties to Strangers and Enemies in a World of 'Dislocated Communities'*. Oxford: Oxford University Press.

Evangelista, Matthew. 2014. 'Introduction'. In *The American Way of Bombing: Changing Ethical and Legal Norms, from Flying Fortresses to Drones*, edited by M. Evangelista and H. Shue. Ithaca, NY: Cornell University Press.

Evans, Michael. 1999. 'Dark Victory'. *United States Naval Institute Proceedings* 125 (9): 33–7.

Evans, Michael. 2017. 'CIA Wins Back Control of Drone Strikes'. Accessed 11 January 2019. https://www.thetimes.co.uk/article/cia-wins-back-control-of-drone-strikes-gvpqf6qqp.

Fabre, Cécile. 2014. *Cosmopolitan War*. Oxford: Oxford University Press.

Farrar, L. L. 1972. 'The Short War Illusion: The Syndrome of German Strategy, August–December 1914'. *Militaergeschictliche Mitteilungen* 2: 39–52.

Farrell, Theo. 2005. *The Norms of War: Cultural Beliefs and Modern Conflict*. Boulder, CO: Lynne Rienner.

Farrell, Theo. 2017. *Unwinnable: Britain's War in Afghanistan, 2001–2014*. London: Bodley Head.

Faust, Drew Gilpin. 2009. *This Republic of Suffering: Death and the American Civil War*. New York: Vintage Books.

Feaver, Peter and Charles Miller. 2014. 'Provocations on Policymakers, Casualty Aversion and Post-Heroic Warfare'. In *Heroism and the Changing Character of War: Toward Post-Heroic Warfare?*, edited by S. Scheipers. Basingstoke: Palgrave Macmillan.

Fierke, Karin M. 2013. *Political Self-Sacrifice: Agency, Body and Emotion in International Relations*. New York: Cambridge University Press.

Fife, Robert. 2017. 'Canadian Elite Special Forces Sniper Makes Record-Breaking Kill Shot in Iraq'. Accessed 11 August 2017. https://beta.theglobeandmail.com/news/politics/canadian-elite-special-forces-sniper-sets-record-breaking-kill-shot-in-iraq/article35415651/?ref=http://www.theglobeandmail.com&.

Filkins, Dexter. 2011. 'The Journalist and the Spies: The Murder of a Reporter who Exposes Pakistan's Secrets'. Accessed 11 December. http://www.newyorker.com/reporting/2011/09/19/110919fa_fact_filkins?currentPage=1.

Finkelstein, Claire. 2012. 'Targeted Killing as Preemptive Action'. In *Targeted Killings: Law and Morality in an Asymmetrical World*, edited by J. D. Ohlin, A. Altman, and C. Finkelstein. Oxford: Oxford University Press.

Finlay, C. J. 2010. 'Legitimacy and Non-State Political Violence'. *The Journal of Political Philosophy* 18 (3): 287–312.

Finnemore, Martha and Kathryn Sikkink. 1998. 'International Norm Dynamics and Political Change'. *International Organization* 52 (4): 887–917.

Fitzsimonds, James and Thomas G. Mahnken. 2007. 'Military Officer Attitudes Toward UAV Adoption'. *Joint Force Quarterly* 46: 96–103.

Flagg, Edmund. 1853. *Venice: The City of the Sea*. Vol. 2. New York: Charles Scribner.

Fleischauer, Eric. 2006. 'Unleashing the Robots of War: Redstone's Robotics Chief Strives to Put Technology, Not Soldiers, in Harm's Way'. Accessed 20 June 2012. http://archive.decaturdaily.com/decaturdaily/news/060619/robots.shtml.

Forbes, Archibald. 1895. *Memories and Studies of War and Peace*. Vol. 1. Leipzig: Bernhard Tauchnitz.

Ford, John C. 1944. 'The Morality of Obliteration Bombing'. *Theological Studies* 5 (3): 261–309.

Fox News. 2001. 'U.S. Kills Al Qaeda Leaders by Remote Control'. Accessed 11 October 2015. http://www.foxnews.com/story/2001/11/19/us-kills-al-qaeda-leaders-by-remote-control.html.

Fox News. 2013. 'America Live | U.S. Military Stops Production of Drone Pilot Medal, 12 MAR 2013'. Accessed 12 October 2015. https://www.youtube.com/watch?v=JvWNrKuSRWE.

Fredrickson, George M. 1968. *The Inner Civil War: Northern Intellectuals and the Crisis of the Union*. Urbana, IL: University of Illinois Press.

Freedman, Lawrence. 2016. 'The Drone Revolution'. *Foreign Affairs* 95 (6): 153–8.

Freire, J. P. 2010. 'NATO Rewarding "Courageous Restraint" Awards'. Accessed 5 March 2016. http://www.washingtonexaminer.com/nato-rewarding-courageous-restraint-awards/article/3997.

French, David. 2015. 'American Sniper Has Created a Cultural Moment: Here's Why'. Accessed 4 April 2019. https://www.nationalreview.com/corner/american-sniper-has-created-cultural-moment-heres-why-david-french/.

French, Shannon E. 2003. *The Code of the Warrior: Exploring Warrior Values Past and Present*. Oxford: Rowman & Littlefield.

Freud, Sigmund. 1961. *Civilization and its Discontents*. New York: W. W. Norton.

Fricker, Matthew, Avery Plaw, and Brian G. Williams. 2010. 'New Light on the Accuracy of the CIA's Predator Drone Campaign in Pakistan'. *Terrorism Monitor* 8 (41): 8–13.

Frontline. 1996. 'The Gulf War'. Accessed 25 May 2013. http://www.pbs.org/wgbh/pages/frontline/gulf/weapons/drones.html.

Frowe, Helen. 2014. *Defensive Killing*. Oxford: Oxford University Press.

Fuhrmann, Matthew and Michael C. Horowitz. 2017. 'Droning On: Explaining the Proliferation of Unmanned Aerial Vehicles'. *International Organization* 71: 397–418.

Fukuyama, Francis. 1989. 'The End of History?' Accessed 14 January 2019. https://www.embl.de/aboutus/science_society/discussion/discussion_2006/ref1-22june06.pdf.

Fulgosius, Raphaël. 2006. 'Just War Reduced to Public War'. In *The Ethics of War*, edited by H. Syse, E. Begby, and G. M. Reichberg. Oxford: Blackwell.

Fuller, J. F. C. 1923. *The Reformation of War*. London: Hutchinson & Co.

Fullerton, J. D. 1894. 'Some Remarks on Aerial Warfare'. In *U.S. Congress 53rd Cong. 2nd Session, Senate Executive Doc 119, Operations of the Division of Military Engineering of the International Congress of Engineers*. Washington, DC.

Fullinwider, Robert K. 1975. 'War and Innocence'. *Philosophy & Public Affairs* 5 (1): 90–7.

Fullinwider, Robert K. 2003. 'Terrorism, Innocence, and War'. In *War After September 11*, edited by V. V. Gehring. Lanham, MD: Rowman & Littlefield.

Furgurson, Ernest B. 2001. *Not War But Murder: Cold Harbor 1864*. New York: Knopf.

Gaita, Raymond. 2003. 'A Last Resort'. In *Why the War Was Wrong*, edited by R. Gaita. Melbourne: The Text Publishing Company.

Galliot, Jai C. 2012. 'Uninhabited Aerial Vehicles and the Asymmetry Objection: A Response to Strawser'. *Journal of Military Ethics* 11 (1): 58–66.

Gamble, C. F. S. 1931. *The Air Weapon*. Vol. 1. Oxford: Oxford University Press.

Garchik, Leah. 1991. 'War Was Healthy and So are Sardines'. *San Francisco Chronicle*, 18 March, A8.

Garlan, Yvon. 1975. *War in the Ancient World: A Social History*. Translated by J. Lloyd. London: Chatto & Windus.

Garraway, Charles. 2014. 'The Law Applies, But which Law?' In *The American Way of Bombing: Changing Ethical and Legal Norms, from Flying Fortresses to Drones*, edited by M. Evangelista and H. Shue. Ithaca, NY: Cornell University Press.

Gellman, Barton. 2001. 'CIA Weighs "Targeted Killing" Missions: Administration Believes Restraints Do Not Bar Singling Out Individual Terrorists'. *Washington Post*, 28 October, A01.

Gelpi, Christopher, Peter D. Feaver, and Jason Reifler. 2005. 'Success Matters: Casualty Sensitivity and the War in Iraq'. *International Security* 30 (3): 7–46.

Gentili, Alberico. 1933. *The Three Books of the Laws of War [De Lure Belli Libri Tres. 1612 ed.]*. Translated by J. C. Rolfe. Oxford: Clarendon Press.

Gentili, Alberico. 2006. 'The Advantages of Preventative War'. In *The Ethics of War*, edited by H. Syse, E. Begby, and G. M. Reichberg. Oxford: Blackwell.

Gettle, Mitch. 2007. 'Air Force Fosters "Warrior Ethos" in all Airmen'. Accessed 10 March 2017. http://ufdcimages.uflib.ufl.edu/AA/00/06/15/90/00016/03-30-2007.pdf.

Gholiagha, Sassan. 2015. 'Individualised and Yet Dehumanised'. *Behemoth: A Journal of Civilisation* 8 (2): 128–53.

Gibbs-Smith, Charles Harvard. 1970. *Aviation: An Historical Survey from its Origins to the End of World War II*. London: HMSO.

Gibson, James. 1986. *The Perfect War: Technowar in Vietnam*. Boston, MA and New York: The Atlantic Monthly Press.

Gibson, Trent. 2009. 'Hell-Bent on Force Protection: Confusing Troop Welfare with Mission Accomplishment in Counterinsurgency'. Master of Military Studies, Marine Corps Combat Development Command, United States Marine Corps, Command and Staff College (AY 08–09).

Gilbert, Adrian. 1997. *Stalk and Kill: The Sniper Experience*. New York: St. Martin's Press.

Gilli, Andrew and Mauro Gilli. 2016. 'So What if Iranian Drones did Strike Syria? We are Not Entering a Dark Age of Robotic Warfare'. Accessed 18 April 2016. https://www.washingtonpost.com/news/monkey-cage/wp/2016/04/04/so-what-if-iranian-drones-did-strike-syria-we-are-not-entering-a-dark-age-of-robotic-warfare/.

Glover, Michael. 1974. *The Peninsular War, 1807–1814: A Concise Military History*. Newton Abbot: David & Charles.

Glover, Michael. 2001. *Humanity: A Moral History of the Twentieth Century*. London: Pimlico.

Glubb, J. B. 1926. 'Air and Ground Force in Punitive Expeditions'. *Journal of Royal United Services Institute* 71 (484): 777–84.

Goldgeier, James M. and Michael McFaul. 1992. 'A Tale of Two Worlds: Core and Periphery in the Post-Cold War era'. *International Organization* 46 (2): 467–91.

Goodman, Ryan. 2013. 'The Power to Kill or Capture Enemy Combatants'. *The European Journal of International Law* 24 (3): 819–53.

Graham, Bradley and William Drozdiak. 1999. '2 Apache Pilots Killed as NATO Raises Tempo'. *Washington Post*, 5 May, A1.

Graves, Robert. 2000. *Goodbye to All That*. London: Penguin.

Gray, Chris H. 1997. *Postmodern War: The New Politics of Conflict*. London: Routledge.

Gray, Chris H. 2001. *Cyborg Citizen: Politics in the Posthuman Age*. New York and London: Routledge.

Gray, Jesse Glenn. 1998. *The Warriors: Reflections on Men in Battle*. Lincoln, NE: University of Nebraska Press.

Green, Leslie C. 1998. 'The Law of War in Historical Perspective'. In *The Law of Military Operations: Liber Amicorum Professor Jack Grunawalt*, edited by M. N. Schmitt. Newport, RI: Naval War College.

Green, Leslie C. 2000. *The Contemporary Law of Armed Conflict*. Manchester: Manchester University Press.

Gregory, Derek. 2016. 'How a Hollywood Film Reveals the Reality of Drone Warfare'. Accessed 11 October 2016. https://www.theguardian.com/world/2016/apr/10/drone-warfare-eye-in-the-sky.

Greppi, Edoardo. 1999. 'The Evolution of Individual Criminal Responsibility Under International Law'. *International Review of the Red Cross* 81 (835): 531–53.

Gross, Michael L. 2006. 'Assassination and Targeted Killing: Law Enforcement, Execution or Self-Defence?' *Journal of Applied Philosophy* 23 (3): 323–35.

Gross, Michael L. 2010. *Moral Dilemmas of Modern War: Torture, Assassination, and Blackmail in an Age of Asymmetrical Conflict*. Cambridge and New York: Cambridge University Press.

Grossman, Dave. 1995. *On Killing: The Psychological Cost of Learning to Kill in War and Society*. Boston, MA: Little, Brown and Company.

Grotius, Hugo. 2012. *On the Law of War and Peace*. Cambridge: Cambridge University Press.

Gusterson, Hugh. 2014. 'Toward an Anthropology of Drones'. In *The American Way of Bombing: Changing Ethical and Legal Norms, from Flying Fortresses to Drones*, edited by M. Evangelista and H. Shue. Ithaca, NY: Cornell University Press.

Gusterson, Hugh. 2016. *Drone: Remote Control Warfare*. Cambridge, MA: MIT Press.

Hacker, Barton C. 2006. 'The Machines of War: Western Military Technology 1850–2000'. *History and Technology: An International Journal* 21 (3): 255–300.

Hadden, Isaac. 1864. 'Issac Hadden to Brother, Wife and All, June 5, 1864, and June 12 1864, Misc'. In *Fallen Leaves: The Civil War Letters of Major Henry Livermore Abbott*, edited by R. G. Scott. Kent, OH: The Kent State University Press.

Haggenmacher, Peter. 1992. 'Just War and Regular War in Sixteenth Century Spanish Doctrine'. *International Review of the Red Cross* 32 (290): 434–45.

Hallgarth, Matthew. 2013. 'Just War Theory and Remote Military Technology: A Primer'. In *Killing by Remote Control: The Ethics of an Unmanned Military*, edited by B. J. Strawser. Oxford: Oxford University Press.

Hallion, Richard. 2003. *Taking Flight: Inventing the Aerial Age, from Antiquity Through the First World War*. New York and Oxford: Oxford University Press.

Hallissy, Margaret. 1987. *Venomous Woman: Fear of the Female in Literature*. New York: Greenwood Press.

Hanson, Victor Davis. 1989. *The Western Way of War: Infantry Battle in Classical Greece*. New York: Oxford University Press.

Hanson, Victor Davis. 1992. *Hoplites: The Classical Greek Battle Experience*. London: Routledge.

Hanson, Victor Davis. 2001. *Carnage and Culture: Landmark Battles in the Rise to Western Power*. New York: Anchor.

Harper, S. J. 2009. 'Ethics versus Morality: A Problematic Divide'. *Philosophy and Social Criticism* 35: 1063–77.

Harris, William. 1979. *War and Imperialism in Republican Rome: 327–70 BC*. Oxford: Clarendon Press.

Hart, B. H. Liddell. 1925. *Paris, Or the Future of War*. New York: E. B. Dutton & Co.

Hart, B. H. Liddell. 1954. *Strategy*. New York: Praeger.

Hartig, Luke. 2017. 'Trump's New Drone Strike Policy: What's Any Different? Why it Matters'. Accessed 24 September 2017. https://www.justsecurity.org/45227/trumps-drone-strike-policy-different-matters/.

Hartigan, Richard S. 1982. *The Forgotten Victim: A History of the Civilian*. Chicago, IL: Precedent Publishing.

Hartog, François. 1988. *The Mirror of Herodotus: The Representation of the Other in the Writing of History*. Translated by J. Lloyd. Berkeley, CA: University of California Press.

Hastings, Max. 2013. *Catastrophe: Europe Goes to War in 1914*. London: William Collins.

Hatto, Arthur Thomas. 1940. 'Archery and Chivalry: A Noble Prejudice'. *The Modern Language Review* 35 (1): 40–54.

Havel, Vaclav. 1999. Speech to a Joint Session of the Senate and House of Commons, Ottawa, 30 April.

Haydon, F. S. 2000. *Military Ballooning During the Early Civil War*. Baltimore, MD: Johns Hopkins University Press.

Hedges, Chris. 2015. 'Killing Ragheads for Jesus'. Accessed 2 April 2015. https://www.truthdig.com/articles/killing-ragheads-for-jesus/.

Hegarty, Stephanie. 2012. 'What Goes On in the Mind of a Sniper?' Accessed 12 March 2015. http://www.bbc.co.uk/news/magazine-16544490.

Heidegger, Martin. 1982. 'The Question Concerning Technology'. In Heidegger, *The Question Concerning Technology, and Other Essays*. New York: Harper Torchbooks.

Hesketh-Prichard, Hesketh. 1920. *Sniping in France, with Notes on the Scientific Training of Scouts, Observers, and Snipers*. New York: E. P. Dutton & Co.

Hewlett, Maurice. 1917. *Our Flying Men*. London: T. B. Hart.

Hibbert, Christopher. 2008. *Redcoats and Rebels*. Barnsley: Pen and Sword.

Higgins, Michael. 2013. 'How a 36-Year-Old Built al-Qaeda in the Arabian Peninsula into a Feared Terrorist Group in Just a Few Short Years'. Accessed 13 May 2017. http://news.nationalpost.com/news/world/israel-middle-east/how-a-36-year-old-managed-to-build-al-qaeda-in-the-arabian-peninsula-into-a-feared-terrorist-group.

Hirsh, Michael. 2011. 'Slow Dance: Obama's Romance with the CIA'. Accessed 10 November 2016. http://www.theatlantic.com/politics/archive/2011/05/slow-dance-obamas-romance-with-the-cia/238849/.

Hobson, John M. 2002. 'What's at Stake in "Bringing Historical Sociology Back into International Relations"? Transcending "Chronofetishism" and "Tempocentrism" in International Relations'. In *Historical Sociology of International Relations*, edited by S. Hobden and J. M. Hobson. Cambridge: Cambridge University Press.

Holder, Eric. 2012. 'Attorney General Eric Holder Speaks at Northwestern University School of Law'. Accessed 14 May 2017. https://www.justice.gov/opa/speech/attorney-general-eric-holder-speaks-northwestern-university-school-law.

Holman, Brett. 2009. 'The First Air Bomb: Venice, 15 July 1849'. Accessed 24 July 2019. https://airminded.org/2009/08/22/the-first-air-bomb-venice-15-july-1849/.

Holmes, Robert. 1989. *On War and Morality*. Princeton, NJ: Princeton University Press.

Homer. 2003. *The Iliad*. Translated by P. Jones, E. V. Rieu, and D. C. H. Rieu. London: Penguin.

Hongsheng, Sheng. 2006. 'The Evolution of Law of War'. *Chinese Journal of International Politics* 1 (2): 267–301.

Honneth, Axel. 1995. *The Struggle for Recognition: The Moral Grammar of Social Conflicts*. Cambridge: Polity Press.

Horkheimer, Max and Theodor W. Adorno. 1995. *Dialectic of Enlightenment: Philosophical Fragments*. Translated by J. Cumming. New York: Continuum Books.

Howard, Michael. 1961. *The Franco-Prussian War: The German Invasion of France, 1870–1871*. London: Rupert Hart-Davis.

Howard, Michael. 1977. *War in European History*. Oxford: Oxford University Press.

Howard, Michael. 1979. 'Temperamenta Belli: Can War be Controlled?' In *Restraints on War: Studies in the Limitation of Armed Conflict*, edited by M. Howard. Oxford: Oxford University Press.

Howard, Michael. 1989. *War and the Liberal Conscience*. Oxford: Oxford University Press.

Howard, Michael. 1994. 'Constraints on Warfare'. In *The Laws of War: Constraints on Warfare in the Western World*, edited by G. J. Andreopoulos, M. R. Shulman, and M. Howard. New Haven, CT: Yale University Press.

Hull, Isabel V. 2005. *Absolute Destruction: Military Culture and the Practices of War in Imperial Germany*. Ithaca, NY: Cornell University Press.

Human Rights Clinic. 2012. 'The Civilian Impact of Drones: Unexamined Costs, Unanswered Questions'. Accessed 12 May 2017. https://www.propublica.org/documents/item/553588-the-civilian-impact-of-drones.

Human Rights Watch. 2012. 'Losing Humanity: The Case Against Killer Robots'. Accessed 14 July 2017. https://www.hrw.org/report/2012/11/19/losing-humanity/case-against-killer-robots.

Human Rights Watch. 2013. 'The Need for New Law to Ban Fully Autonomous Weapons: Memorandum to Convention on Conventional Weapons Delegates'. Accessed 2 September 2014. https://www.hrw.org/sites/default/files/supporting_resources/11.2013_memo_to_ccw_delegates_fully_autonomous_weapons.pdf

Hurka, Thomas. 2007. 'Liability and Just Cause'. *Ethics and International Affairs* 21 (2): 199–218.

Hylton, Stuart. 2014. *Reporting the Great War: News from the Home Front*. Barnsley: Pen and Sword Military.

Ignatieff, Michael. 2000. *Virtual War: Kosovo and Beyond*. New York: Henry Holt.

International Committee of the Red Cross. 1863. 'Instructions for the Government of Armies of the United States in the Field (Lieber Code)'. Accessed 20 April 2013. https://ihl-databases.icrc.org/applic/ihl/ihl.nsf/Treaty.xsp?action=openDocument&documentId=A25AA5871A04919BC12563CD002D65C5.

International Committee of the Red Cross. 1864. 'Convention for the Amelioration of the Condition of the Wounded in Armies in the Field. Geneva, 22 August 1864'. Accessed 11 May 2017. https://ihl-databases.icrc.org/ihl/INTRO/120?OpenDocument.

International Committee of the Red Cross. 1868. 'Declaration Renouncing the Use, in Time of War, of Explosive Projectiles Under 400 Grammes Weight. Saint Petersburg, 29 November / 11 December 1868'. Accessed 11 May 2017. https://ihl-databases.icrc.org/ihl/WebART/130-60001?OpenDocument.

International Committee of the Red Cross. 1880. 'The Laws of War on Land. Oxford, 9 September 1880'. Accessed 10 April 2015. https://ihl-databases.icrc.org/applic/ihl/ihl.nsf/385ec082b509e76c41256739003e636d/6a5d425d29d9d6dbc125641e0032ec97

International Committee of the Red Cross. 1899a. 'Declaration (IV, 3) Concerning Expanding Bullets. The Hague, 29 July 1899'. Accessed 23 December 2018. https://ihl-databases.icrc.org/applic/ihl/ihl.nsf/Article.xsp?action=openDocument&documentId=F5FF4D9CA7E41925C12563CD0051616B.

International Committee of the Red Cross. 1899b. 'Declaration (IV, 2) Concerning Asphyxiating Gases. The Hague, 29 July 1899'. Accessed 23 December 2018. https://ihl-databases.icrc.org/applic/ihl/ihl.nsf/Treaty.xsp?action=openDocument&documentId=B0625F804A9B2A64C12563CD002D66FF.

International Committee of the Red Cross. 1906. 'Convention for the Amelioration of the Condition of the Wounded and Sick in Armies in the Field. Geneva, 6 July 1906'. Accessed 13 May 2017. https://ihl-databases.icrc.org/ihl/INTRO/180?OpenDocument.

International Committee of the Red Cross. 1907a. 'Convention (IV) Respecting the Law and Customs of War on Land and its Annex: Regulations Concerning the Laws and Customs of War on Land. The Hague, 18 October 1907'. Accessed 12 February 2013. https://ihl-databases.icrc.org/ihl/INTRO/195.

International Committee of the Red Cross. 1907b. 'Declaration (XIV) Prohibiting the Discharge of Projectiles and Explosive from Balloons. The Hague, 18 October 1907'. Accessed 10 July 2015. https://www.icrc.org/applic/ihl/ihl.nsf/INTRO/245?OpenDocument.

International Committee of the Red Cross. 1923. 'Rules concerning the Control of Wireless Telegraphy in Time of War and Air Warfare. Drafted by a Commission of Jurists at the Hague, December 1922–February 1923'. Accessed 11 April 2015. https://ihl-databases.icrc.org/ihl/INTRO/275

International Committee of the Red Cross. 1938. 'Draft Convention for the Protection of Civilian Populations Against New Engines of War. Amsterdam, 1938'. Accessed 4 August 2015. https://www.icrc.org/applic/ihl/ihl.nsf/0/910f79361f226492c125641e004057ed?OpenDocument.

International Committee of the Red Cross. 1949. 'Convention (IV) Relative to the Protection of Civilian Persons in Time of War. Geneva, 12 August 1949'. Accessed 11 November 2016. https://ihl-databases.icrc.org/applic/ihl/ihl.nsf/Treaty.xsp?documentId=AE2D398352C5B028C12563CD002D6B5C&action=openDocument.

International Committee of the Red Cross. 1950. 'Principles of International Law Recognised in the Charter of the Nüremburg Tribunal and in the Judgment of the Tribunal, 1950'. Accessed 17 March 2017. https://ihl-databases.icrc.org/applic/ihl/ihl.nsf/Treaty.xsp?action=openDocument&documentId=854DDAACFDE285E4C12563CD002D6B95.

International Committee of the Red Cross. 1977. 'Protocol Additional to the Geneva Conventions of 12 August 1949, and relating to the Protection of Victims of International Armed Conflicts (Protocol I), 8 June 1977'. Accessed 23 March 2013. http://www.icrc.org/ihl.nsf/full/470?opendocument.

International Committee of the Red Cross. 1998. 'Rome Statute of the International Criminal Court, 17 July 1998'. Accessed 10 December 2015. https://www.icrc.org/ihl/INTRO/585?OpenDocument.

International Committee of the Red Cross. 2008. 'How is the Term "Armed Conflict" Defined in International Humanitarian Law?' Accessed 20 January 2019. https://www.icrc.org/en/doc/assets/files/other/opinion-paper-armed-conflict.pdf.

International Committee of the Red Cross. 2017. 'Practice Relating to Rule 89. Violence to Life'. Accessed 10 May 2017. https://ihl-databases.icrc.org/customary-ihl/eng/docs/v2_cha_chapter32_rule89?OpenDocument&highlight=sniping.

Jackson, Robert. 2000. *The Global Covenant: Human Conduct in a World of States*. Oxford and New York: Oxford University Press.

Johansson, Linda. 2011. 'Is it Morally Right to Use Unmanned Aerial Vehicles (UAVs) in War?' *Philosophy and Technology* 24: 279–91.

Johns, Elizabeth. 2002. *Winslow Homer: The Nature of Observation*. London: University of California Press.

Johnson, James Turner. 1981. *Just War Tradition and the Restraint of War: A Moral and Historical Inquiry*. Princeton, NJ: Princeton University Press.

Johnson, James Turner. 1999. *Morality and Contemporary Warfare*. New Haven, CT: Yale University Press.

Johnson, James Turner. 2006. 'Maintaining the Protection of Non-Combatants'. In *The Ethics of War: Classical and Contemporary Readings*, edited by H. Syse, E. Begby, and G. M. Reichberg. Malden, MA: Blackwell Publishing.

Johnson, James Turner. 2011. *Ethics and the Use of Force: Just War in Historical Perspective*. Farnham: Ashgate.

Johnson, Jenna. 2016. 'Donald Trump on Waterboarding: "Torture Works"'. Accessed 11 July 2017. https://www.washingtonpost.com/news/post-politics/wp/2016/02/17/donald-trump-on-waterboarding-torture-works/.

Johnston, Alan. 2011. 'Libya 1911: How an Italian Pilot Began the Air War Era'. Accessed 20 May 2015. http://www.bbc.co.uk/news/world-europe-13294524.

Johnston, David and David E. Sanger. 2002. 'Threats and Responses: Hunt for Suspects; Fatal Strike in Yemen Was Based on Rules Set Out by Bush'. Accessed 4 June 2017. http://www.nytimes.com/2002/11/06/world/threats-responses-hunt-for-suspects-fatal-strike-yemen-was-based-rules-set-bush.html.

Jünger, Ernst. 1929. *Storm of Steel*. Translated by B. Creighton. London: Chatto & Windus.

Jünger, Ernst. 2004. *Storm of Steel*. Translated by M. Hofmann. New York: Penguin Books.

Junger, Sebastian. 2010. *War*. London: Fourth Estate.

Kaag, John and Sarah Kreps. 2014. *Drone Warfare*. Cambridge: Polity Press.

Kaempf, Sebastian. 2009. 'Double Standards in US Warfare: Exploring the Historical Legacy of Civilian Protection and the Complex Nature of the Moral–Legal Nexus'. *Review of International Studies* 35: 651–74.

Kaempf, Sebastian. 2014. 'Post-Heroic US Warfare and the Moral Justification for Killing in War'. In *The Future of Just War*, edited by C. Gentry and A. Eckert. Athens, GA: Georgia University Press.

Kahn, Paul. 1999. 'War and Sacrifice in Kosovo'. *Philosophical Dimensions of Public Policy* 13: 201.

Kahn, Paul. 2002. 'The Paradox of Riskless Warfare'. *Philosophy & Public Policy Quarterly* 22 (3): 2–7.

Kaldor, Leif and Leslea Mair. 2011. *Remote Control War*. Film. PBS.

Kalmanovitz, Pablo. 2015. 'Early Modern Sources of the Regular War Tradition'. In *The Oxford Handbook of Ethics of War*, edited by S. Lazar and H. Frowe. Oxford: Oxford University Press.

Kant, Immanuel. 1954. 'What is Enlightenment?'. In *Introduction to Contemporary Civilisation in the West*, edited by P. Gay. New York: Columbia University Press.

Karon, Tony. 2002. 'Yemen Strike Opens New Chapter in War on Terror'. Accessed 13 May 2017. http://content.time.com/time/world/article/0,8599,387571,00.html.

Kasher, Asa. 2014. 'Combatants' Life and Human Dignity'. *Israel Yearbook on Human Rights* 24: 219–46.

Kasher, Asa and Amos Yadlin. 2005. 'Military Ethics of Fighting Terror: An Israeli Perspective'. *Journal of Military Ethics* 4 (1): 3–32.

Katcher, Philip. 2002. *Sharpshooters of the American Civil War 1861–65*. Vol. 60. Oxford: Osprey Publishing.

Kaurin, Pauline M. 2016. *The Warrior, Military Ethics and Contemporary Warfare: Achilles Goes Asymmetrical*. London: Routledge.

Keane, John and S. Carr. 2013. 'A Brief History of Early Unmanned Aircraft'. *Johns Hopkins APL Technical Digest* 32 (3): 558–71.

Keegan, John. 1994. *A History of War*. London: Pimlico.

Keen, Maurice. 1984. *Chivalry*. New Haven, CT: Yale University Press.

Kellenberger, Jakob and Philip Spoerri. 2011. 'International Humanitarian Law and New Weapon Technologies, 34th Round Table on Current Issues of International Humanitarian Law, San Remo, 8-10 September 2011'. Accessed 10 October 2016. https://international-review.icrc.org/sites/default/files/irrc-886-kellenberger-spoerri.pdf

Kels, Charles. 2012. 'Closing Argument: At the Outer Bounds of Asymmetry'. *Journal of Military Ethics* 11 (3): 223–44.

Kels, Charles. 2014. 'Guest Post: Henderson and Cavanagh on Self-Defense During Armed Conflict'. Accessed 12 September 2015. http://opiniojuris.org/2014/07/25/guest-post-henderson-cavanagh-self-defense-armed-conflict/.

Kennedy, David. 2012. 'Lawfare and Warfare'. In *The Cambridge Companion to International Law*, edited by J. Crawford and M. Koskenniemi. Cambridge: Cambridge University Press.

Kennett, Lee. 1991. *The First Air War*. New York: Free Press.

Kenny, Anthony. 1985. *The Logic of Deterrence*. London: University of Chicago Press.

Kershnar, Stephen. 2005. 'Assassination and the Immunity Theory'. *Philosophia* 33 (1–4): 129–47.

Khan, Azmat and Anand Gopal. 2017. 'The Uncounted'. Accessed 11 February 2019. https://www.nytimes.com/interactive/2017/11/16/magazine/uncounted-civilian-casualties-iraq-airstrikes.html?_r=0.

Kilcullen, David and Andrew Exum. 2009. 'Death from Above, Outrage Down Below'. *The New York Times*, 16 May.

Killmister, Suzy. 2008. 'Remote Weaponry: The Ethical Implications'. *Journal of Applied Philosophy* 25 (2): 121–33.

Kirkpatrick, Jesse. 2015. 'Drones and the Martial Virtue Courage'. *Journal of Military Ethics* 14 (3–4): 202–19.

Kirkpatrick, Jesse. 2016. 'Military Drone Operators Risk a Serious Injury'. Accessed 15 May 2017. http://www.slate.com/articles/technology/future_tense/2016/01/military_drone_operators_risk_moral_injury.html.

Klaidman, Daniel. 2012. *Kill or Capture: The War on Terror and the Soul of the Obama Presidency*. New York: Houghton Mifflin.

Kramer, Michael and Michael Schmitt. 2008. 'Lawyers on Horseback? Thoughts on Judge Advocates and Civil–Military Relations'. *UCLA Law Review* 55 (5): 1407–36.

Krasner, Jeffrey. 2007. 'Robots Going in Harm's War: Agile New Devices Save Lives by Disarming or Detonating Roadside Bombs'. *Boston Globe*, 12 March.

Krentz, Peter. 2000. 'Deception in Archaic and Classical Warfare'. In *War and Violence in Ancient Greece*, edited by H. V. Wees and P. Beston. London: Duckworth and the Classical Press of Wales.

Kurtz, Howard. 1999. 'Media Notes: Airstrikes Followed by Retreat'. *Washington Post*, 5 April.

Kwek, Glenda. 2015. 'Chinese War Hero Australia's Top Gallipoli Sniper'. Accessed 14 May 2015. http://www.smh.com.au/world/chinese-war-hero-australias-top-gallipoli-sniper-20150422-1mqem8.html.

Lackey, Douglas. 2008. 'Pacifism'. In *Contemporary Moral Problems*, edited by J. E. White. Belmont, CA and London: Wadsworth Publishing.

Lacquement, Richard A. 2004. 'The Casualty-Aversion Myth'. *Naval War College Review* 57 (1): 38–57.

LaGrone, Sam. 2015. 'NAVAIR: Aerial Refuelling Will End X-47B Test Program, Salty Dogs Bound for Museums or Boneyard'. Accessed 5 April 2016. https://news.usni.org/2015/04/14/navair-aerial-refueling-will-end-x-47b-test-program-salty-dogs-bound-for-museums-or-boneyard.

Lamb, Antony. 2013. *Ethics and the Laws of War: The Moral Justification of Legal Norms*. Abingdon: Routledge.

Lambeth, Benjamin S. 2000. *The Transformation of American Air Power*. Ithaca, NY: Cornell University Press.

Lamp, Nicolas. 2011. 'Conceptions of War and Paradigms of Compliance: The "New War" Challenge to International Humanitarian Law'. *Journal of Conflict and Security Law* 16 (2): 225–62.

Lance, David. 1976. 'Folklore of Aviation'. *The RUSI Journal* 121 (1): 62–8.

Landay, Jonathan S. 2013. 'Obama's Drone War Kills "Others," Not Just Al-Qaida Leaders'. Accessed 6 January 2016. https://www.mcclatchydc.com/news/nation-world/world/article24747826.html.

Lanni, Adriaan. 2008. 'The Laws of War in Ancient Greece'. *Law and History Review* 26 (3): 469–89.

Larter, David B. 2017. 'SOCOM Commander: Armed ISIS Drones Were 2016's "Most Daunting Problem"'. Accessed 4 June 2017. http://www.defensenews.com/articles/socom-commander-says-isis-drones-were-2016s-most-daunting-problem.

'Last Word on Telechirics'. 1964. *New Scientist* 22 (391): 405.

Latimer, Jon. 1998. 'Armed with Lethal New Sharps and Whitworth Rifles, Sharpshooters Could Kill from a Mile Away'. *America's Civil War* 11 (5): 12–16.

Lattimore, Richmond. 1955. *Greek Lyrics*. Chicago, IL: University of Chicago Press.

Lauterbach, Paul. 1915. 'Annex C-56'. In *The Belgian People's War: A Violation of International Law*. New York: Press of John C. Rankin Co.

Lauterpacht, Hersch. 1952. *L. F. L Oppenheim. International Law: A Treatise*. Vol. 2. London and New York: Longmans, Green and Co.

Lauterpacht, Hersch. 2004. 'The Limits of the Operation of the Laws of War'. In *International Law: Collected Papers*, edited by E. Lauterpacht. Cambridge: Cambridge University Press.

Lawson, George. 2012. 'The Eternal Divide? History and International Relations'. *European Journal of International Relations* 18 (2): 203–26.

Lawson, George and John M. Hobson. 2008. 'What is History in International Relations?' *Millennium: Journal of International Studies* 37 (2): 415–35.

Lazar, Seth. 2010. 'The Responsibility Dilemma for Killing in War: A Review Essay'. *Philosophy & Public Affairs* 38 (2): 180–213.

Lazar, Seth. 2015. *Sparing Civilians*. New York: Oxford University Press.

League of Nations. 1932. 'Conference for the Reduction and Limitation of Armaments'. Accessed 12 August 2015. http://digital.library.northwestern.edu/league/le000007.pdf.

Lee, Peter. 2012. 'Remoteness, Risk and Aircrew Ethos'. *Air Power Review* 15 (1): 1–19.

Lee, Peter. 2014. 'Exploring the Roles of Personal Ethics, Individual Identity and Operational Practices in the Formation of a Collective Ethos in RAF Reaper Squadrons'. MOD Research Ethics Committee Protocol No. 423/MODREC/13.

Lewis, Cecil. 1936. *Sagittarius Rising*. London: P. Davies.

Lichtenberg, Judith. 2008. 'How to Judge Soldiers Whose Cause Is Unjust?' In *Just and Unjust Warriors: The Moral and Legal Status of Soldiers*, edited by D. Rodin and H. Shue. Oxford: Oxford University Press.

Lieber, Francis. 1862. 'The Usages of War; Continuation of Dr. Lieber's Lecture – The Treatment of the Wounded – of the Physicians – Firing on Sentinels – Flags of Truce in War – Treatment of the Spy'. Accessed 11 April 2015. http://www.nytimes.com/1862/02/04/news/usages-war-continuation-dr-lieber-s-lecture-treatment-wounded-physicians-firing.html.

Lieber, Francis. 1889. *Manual of Political Ethics, Part 2: Political Ethics Proper (for the use of colleges and students in law)*. Boston, MA: Charles C. Little and James Brown. Accessed 13 March 2018. https://psycnet.apa.org/PsycBOOKS/toc/14235

Lin, Patrick, George Bekey, and Keith Abney. 2008. 'Autonomous Military Robotics: Risk, Ethics, and Design'. Accessed 12 September 2016. http://ethics.calpoly.edu/onr_report.pdf.

Linden, Harry van der. 2015. 'Drone Warfare and Just War Theory'. In *Drones and Targeted Killing: Legal, Moral, and Geopolitical Issues*, edited by M. Cohn. Northampton, MA: Olive Branch Press.

Lindqvist, Sven. 2001. *A History of Bombing*. London: Granta.

Linklater, Andrew. 2002. 'The Problem of Harm in World Politics: Implications for the Sociology of State-Systems'. *International Affairs* 78 (2): 319–38.

Llenza, Steven. 2011. 'Targeted Killings in Pakistan: A Defence'. *Global Security Studies* 2 (2): 47–59.

LoBianco, Tom. 2015. 'Donald Trump on Terrorists: "Take out their Families"'. Accessed 11 July 2017. http://edition.cnn.com/2015/12/02/politics/donald-trump-terrorists-families/.

Longden, Sean. 2007. *To the Victor the Spoils: Soldiers' Lives from D-Day to VE-Day*. London: Robinson.

Lorenz, Konrad Z. 2010. 'On Aggression'. In *Approaches to Peace: A Reader in Peace Studies*, edited by D. P. Barash. New York and Oxford: Oxford University Press.

Löwith, Karl. 1989. 'Max Weber's Position on Science'. In *Max Weber's 'Science as a Vocation'*, edited by I. Velody, H. Martins, and P. Lassman. London: Unwin Hyman.

Lubell, N. 2010. *Extraterritorial Use of Force Against Non-State Actors*. Oxford: Oxford University Press.

Luttwak, Edward N. 1995. 'Toward Post-Heroic Warfare'. *Foreign Affairs* 74: 109–22.

Luttwak, Edward N. 1999. 'Give War a Chance'. *Foreign Affairs* 78 (4): 36–44.

Lyon, Charlie. 2001. 'Operation Allied Force: A Lesson on Strategy, Risk, and Tactical Execution'. *Comparative Strategy* 20 (1): 57–75.

McCartney, Helen. 2010. 'The Military Covenant and the Civil–Military Contract in Britain'. *International Affairs* 86 (2): 411–28.

McCloskey, Megan. 2009. 'The War Room: Daily Transition Between Battle, Home Takes a Toll on Drone Operators'. Accessed 9 November 2015. http://www.stripes.com/news/the-war-room-daily-transition-between-battle-home-takes-a-toll-on-drone-operators-1.95949.

McCudden, James. 2009. *Flying Fury*. Philadelphia, PA: Casemate.

McInnes, Colin. 2002. *Spectator-Sport War*. Boulder, CO: Lynne Rienner.

McKenzie, Kenneth F. 2000. *The Revenge of the Melians: Asymmetric Threats and the Next QDR*. Washington, DC: Institute for National Strategic Studies.

McKitrick, Jeffrey, James Blackwell, Fred Littlepage, George Kraus, Richard Blanchfield, and Dale Hill. 1998. 'The Revolution in Military Affairs'. In *Battlefield of the Future: 21st Century Warfare Issues*, edited by B. Schneider and L. E. Grinter. Maxwell Air Force Base, AL: Air University Press.

McMahan, Jeff. 2004a. 'Response to War as Self-Defense: War as Self-Defence'. *Ethics & International Affairs* 18 (1): 75–80.

McMahan, Jeff. 2004b. 'The Ethics of Killing in War'. *Ethics* 114 (4): 693–733.

McMahan, Jeff. 2005. 'Just Cause for War'. *Ethics and International Affairs* 19 (3): 1–21.

McMahan, Jeff. 2006. 'Killing in War: A Reply to Walzer'. *Philosophia* 34: 47–51.

McMahan, Jeff. 2008. 'The Morality of War and the Law of War'. In *Just and Unjust Warriors: The Moral and Legal Status of Soldiers*, edited by D. Rodin and H. Shue. Oxford: Oxford University Press.

McMahan, Jeff. 2009. *Killing in War*. Oxford: Oxford University Press.

McMahan, Jeff. 2010. 'Pacifism and Moral Theory'. *Diametros* 23: 44–68.

McMahan, Jeff. 2011. 'Who is Morally Liable to be Killed in War?' *Analysis Reviews* 71 (3): 544–59.

McMahan, Jeff. 2013. 'Foreword'. In *Killing by Remote Control: The Ethics of an Unmanned Military*, edited by B. J. Strawser. Oxford: Oxford University Press.

Macmanus, Joseph E. and Elizabeth L. King. 2011. 'United States Activities in Libya'. Accessed 11 September 2016. https://fas.org/man/eprint/wh-libya.pdf.

MacPherson, Malcolm. 2005. *Robert's Ridge*. London: Bantam Books.

Mailer, Norman. 1999. 'Milosevic and Clinton'. *Washington Post*, 24 May, A25.

Manganiello, Stephen C. 2004. *The Concise Encyclopaedia of the Revolutions and Wars of England, Scotland, and Ireland, 1639–1660*. Oxford: Scarecrow Press.

Marcuse, Herbert. 1991. *One-Dimensional Man: Studies in the Ideology of Advanced Industrial Society*. 2nd edition. London: Routledge.

Marder, Arthur J. 1961. *Fear God and Dread Nought: The Correspondence of Admiral of the Fleet Lord Fisher of Kilverstone*. Vol. 1. London: Jonathan Cape.

Martin, Matt J. and Charles W. Sasser. 2010. *Predator: The Remote Control Air War over Iraq and Afghanistan: A Pilot's Story*. Minneapolis, MN: Zenith Press.

Martin, Rachel. 2011. 'Report: High Levels of "Burnout" in U.S. Drone Pilots'. Accessed 11 November 2015. http://www.npr.org/2011/12/19/143926857/report-high-levels-of-burnout-in-u-s-drone-pilots.

Mason, Tony. 2010. 'Operation Allied Force, 1999'. In *A History of Air Warfare*, edited by J. A. Olsen. Washington, DC: Potomac Books.

Mast, Gregory and Hans Halberstadt. 2007. *To be a Military Sniper*. Minneapolis, MN: Zenith Press.

Mavrodes, George I. 1975. 'Conventions and the Morality of War'. *Philosophy & Public Affairs* 4 (2): 117–31.

May, Larry. 2005. 'Killing Naked Soldiers: Distinguishing between Combatants and Noncombatants'. *Ethics and International Affairs* 19 (3): 39–53.

May, Larry. 2013. 'Targeted Killings and Proportionality in Law: Two Models'. *Journal of International Criminal Justice* 11 (1): 47–63.

May, Larry. 2015a. 'The Nature of War and the Idea of "Cyberwar"'. In *Cyberwar: Law and Ethics for Virtual Conflicts*, edited by K. Govern, C. Finkelstein, and J. O. Ohlin. Oxford: Oxford University Press.

May, Larry. 2015b. *Contingent Pacifism: Revisiting Just War Theory*. Cambridge: Cambridge University Press.

Mayer, Jane. 2009. 'The Predator War: What Are the Risks of the C.I.A.'s Covert Drone Program?' Accessed 11 October 2015. http://www.newyorker.com/magazine/2009/10/26/the-predator-war.

Mayer, Michael. 2015. 'The New Killer Drones: Understanding the Strategic Implications of Next-Generation Unmanned Combat Aerial Vehicles'. *International Affairs* 9 (4): 765–80.

Mazzetti, Mark. 2013. *The Way of the Knife: The CIA, a Secret Army, and a War at the Ends of the Earth*. Melbourne: Scribe.

Meagher, Robert Emmet. 2014. *Killing from the Inside Out: Moral Injury and Just War*. Eugene, OR: Cascade Books.

Meijering, Piet Hein. 1987. *Signed with their Honour: The Story of Chivalry in Air Warfare 1914–45*. Edinburgh: Mainstream.

Meilinger, Phillip. 1993. 'Winged Defence: Airway, the Law, and Morality'. *Armed Forces and Society* 20 (1): 103–23.

Meilinger, Phillip. 2003. *Airpower Myths and Facts*. Maxwell Air Force Base, AL: Air University Press.

Meilinger, Phillip. 2008. 'Disarmament and Airpower, 1932–1934: Clipping the Bomber's Wings'. *The RUSI Journal* 142 (5): 79–82.

Meisels, Tamar. 2007. 'Combatants: Lawful and Unlawful'. *Law and Philosophy* 26 (1): 31–65.

Meisels, Tamar. 2017. *Contemporary Just War: Theory and Practice*. London: Routledge.

Mellor, Ewan E. 2016. 'Just and Unjust Distributions in War: The Moral Equality of Combatants and the Normative Structure of War'. PhD thesis, Department of Political and Social Sciences, European University Institute.

Meron, Theodor. 2006. *The Humanization of International Law*. Leiden: Martinus Nijhoff.

Meyer, Richard V. 2012. 'The Privilege of Belligerency and Formal Declarations of War'. In *Targeted Killings: Law and Morality in an Asymmetrical World*, edited by J. D. Ohlin, A. Altman, and C. Finkelstein. Oxford: Oxford University Press.

Mill, John Stuart. 2008. *Dissertations and Discussions*. Vol. 3. New York: Cosimo Classics.

Miller, Bob. 2007. *Kill Me If You Can, You SOB*. Tucson, AZ: Wheatmark.

Miller, Greg. 2012. 'CIA Seeks New Authority to Expand Yemen Drone Campaign'. Accessed 9 May 2017. https://www.washingtonpost.com/world/national-security/cia-seeks-new-authority-to-expand-yemen-drone-campaign/2012/04/18/gIQAsaumRT_story.html?tid=pm_world_pop.

Miller, Greg. 2015. 'U.S. Launches Secret Drone Campaign to Hunt Islamic State Leaders in Syria'. Accessed 10 November 2016. https://www.washingtonpost.com/world/national-security/us-launches-secret-drone-campaign-to-hunt-islamic-state-leaders-in-syria/2015/09/01/723b3e04-5033-11e5-933e-7d06c647a395_story.html.

Mills, R. 1894. 'The Bond of Heroism'. In *Southern Historical Society Papers, 1876–1959*, edited by R. A. Brock. Millswood, VA: The Society.

Mindell, David A. 2000. *War, Technology, and Experience Aboard the USS Monitor* Baltimore, MD: Johns Hopkins University Press.

Mockenhaupt, Brian. 2009. 'We've Seen the Future, and it's Unmanned'. Accessed 11 November 2015. http://www.esquire.com/news-politics/a6379/unmanned-aircraft-1109/.

Molter, Bennet A. 1918. *Knights of the Air*. New York and London: D. Appleton & Co.

Montague, Phillip. 2012. 'Defending Defensive Targeted Killings'. In *Targeted Killings: Law and Morality in an Asymmetrical World*, edited by J. D. Ohlin, A. Altman, and C. Finkelstein. Oxford: Oxford University Press.

Montluc, Blaise de. 1972. *The Habsburg-Valois Wars and the French Wars of Religion*, edited by I. Roy. Hamden, CT: Archon Books.

Moore, Frank, ed. 1862. *The Rebellion Record: A Diary of American Events*. Vol. 6. New York: G. P. Putnam. Accessed 25 July 2019. https://ia802308.us.archive.org/6/items/rebellrecord06moorrich/rebellrecord06moorrich.pdf

Moore, Michael. 2015. 'Twitter, 18 Jan 2015'. Accessed 20 April 2015. https://twitter.com/mmflint/status/556914094406926336.

Morison, Elting E. 1966. *Men, Machines, and Modern Times*. Cambridge, MA: MIT Press.

Mosher, Dave and Skye Gould. 2017. 'How Likely Are Foreign Terrorists to Kill Americans? The Odds May Surprise You'. Accessed 5 June 2017. https://www.businessinsider.com.au/death-risk-statistics-terrorism-disease-accidents-2017-1?r=US&IR=T.

Mueller, John. 1995. 'The Perfect Enemy: Assessing the Gulf War'. *Security Studies* 5 (1): 77–117.
Mueller, John. 2004. *The Remnants of War*. Ithaca, NY: Cornell University Press.
Mueller, John and Mark G. Stewart. 2012. 'The Terrorism Delusion: America's Overwrought Response to September 11'. *International Security* 37 (1): 81–110.
Münkler, Herfried. 2003. 'The Wars of the 21st Century'. *International Review of the Red Cross* 85: 7–22.
Muñoz-Rojas, D. and J. Frésard. 2004. 'The Roots of Behavior: Understanding and Preventing IHL Violations'. *International Review of the Red Cross* 86 (853): 189–206.
Murphy, Jeffrie. 1973. 'The Killing of the Innocent". *The Monist* 57 (4): 527–50.
Nadelson, Theodore. 2005. *Trained to Kill: Soldiers at War*. Baltimore, MD: Johns Hopkins University Press.
Nagel, Thomas. 1972. 'War and Massacre'. *Philosophy & Public Affairs* 1 (2): 123–44.
National Army Museum. 2015. 'Sniper'. Accessed 11 September 2016. https://collection.nam.ac.uk/detail.php?acc=1986-11-15-10.
Naylor, Sean. 2015. 'Inside the Pentagon's Manhunting Machine: A Brief History of Special Operations, From Panama to the War on Terror'. *The Atlantic*, 28 August.
Nelson, Dave. 1992. 'Dave Nelson'. In *Parallels: The Soldier's Knowledge and the Oral History of Contemporary Warfare*, edited by A. S. Owen, M. Patrick, and J. T. Hansen. New York: Aldine de Gruyter.
Newcome, Laurence R. 2004. *Unmanned Aviation: A Brief History of Unmanned Vehicles*. Barnsley: Pen and Sword Aviation.
Newsweek. 1991. 'A Soldier of Conscience'. Accessed 10 March 2017. http://www.newsweek.com/soldier-conscience-201554.
Newton, Michael A. and Larry May. 2014. *Proportionality in International Law*. New York: Oxford University Press.
Nicholson, Brendan. 2015. 'RAAF "Red Cards" Spare Civilians in Iraq'. Accessed 17 August 2015. https://www.theaustralian.com.au/nation/defence/raaf-red-cards-spare-civilians-in-iraq/news-story/40d3cd57a07ca5921d20ffd555611201
Nietzsche, Friedrich. 1979. *The Portable Nietzsche*. New York: Picador.
Norman, Richard. 1995. *Ethics, Killing and War*. Cambridge: Cambridge University Press.
Nussbaum, Rachel L. 2015. 'Changing the Tooth-to-Tail Ratio Using Robotics and Automation to Beat Sequestration'. *Air and Space Journal* (September–October). Accessed 12 January 2019. https://www.airuniversity.af.edu/Portals/10/ASPJ/journals/Volume-29_Issue-5/V-Nussbaum.pdf
Nyhamar, Tore. 2000. 'How Do Norms Work? A Theoretical and Empirical Analysis of African International Relations'. *International Journal of Peace Studies* 5 (2): 27–43.
O'Connell, Mary E. 2010. 'Declaration of Prof. Mary Ellen O'Connell, Al-Aulaqi v. Obama'. Accessed 11 May 2017. https://ccrjustice.org/sites/default/files/assets/Declaration%20of%20Mary%20Ellen%20O'Connell%2010-08-2010.pdf.
O'Connell, Mary E. 2011. 'Seductive Drones: Learning from a Decade of Lethal Operations'. *Journal of Law, Information and Science* 21 (2): 116–39.
O'Connell, Mary E. 2012. 'Why Obama's "Targeted Killing" Is Worse than Bush's Torture'. Accessed 20 December 2015. http://www.theguardian.com/commentisfree/cifamerica/2012/jan/20/why-obama-targeted-killing-is-like-bush-torture.
O'Connell, Mary E. 2014. 'What's Wrong with Drones? The Battlefield in International Humanitarian Law'. In *The American Way of Bombing: Changing Ethical and Legal Norms, from Flying Fortresses to Drones*, edited by M. Evangelista and H. Shue. Ithaca, NY: Cornell University Press.

O'Connell, Mary E. 2015. 'International Law and Drone Attacks beyond Armed Conflict Zones'. In *Drones and the Future of Conflict: Ethical, Legal, and Strategic Implications*, edited by R. Fairhurst, K. Wall, and D. Cortright. Chicago, IL: University of Chicago Press.

O'Connell, Robert L. 1989. *Of Arms and Men: A History of War, Weapons, and Aggression*. New York: Oxford University Press.

O'Donovan, Oliver. 2003. *The Just War Revisited*. Cambridge: Cambridge University Press.

O'Driscoll, Cian. 2015. 'Rewriting the Just War Tradition: Just War in Classical Greek Political Thought and Practice'. *International Studies Quarterly* 59 (1): 1–10.

O'Hehir, Andrew. 2015. '"American Sniper" and the Culture Wars: Why the Movie's Not What You Think It Is'. Accessed 3 April 2019. https://www.salon.com/2015/01/21/american_sniper_and_the_culture_wars_why_the_movies_not_what_you_think_it_is/.

Obama, Barack. 2013. 'Remarks by the President at the National Defense University'. Accessed 14 November 2016. https://www.whitehouse.gov/the-press-office/2013/05/23/remarks-president-national-defense-university.

Ober, Josiah. 1994. 'Classical Greek Times'. In *The Laws of War: Constraints on Warfare in the Western World*, edited by G. J. Andreopoulos, M. R. Shulman, and M. Howard. New Haven, CT: Yale University Press.

Ohlin, Jens David. 2012. 'Targeting Co-Belligerents'. In *Targeted Killings: Law and Morality in an Asymmetrical World*, edited by J. D. Ohlin, A. Altman, and C. Finkelstein. Oxford: Oxford University Press.

Ohlin, Jens David and Larry May. 2016. *Necessity in International Law*. New York: Oxford University Press.

Olsthoorn, Peter. 2011. *Military Ethics and Virtues: An Interdisciplinary Approach for the 21st Century*. New York: Routledge.

Omissi, David E. 1990. *Air Power and Colonial Control: The Royal Air Force, 1919–1939*. Manchester: Manchester University Press.

Orend, Brian. 2000. *Michael Walzer on War and Justice*. Montreal: McGill-Queen's University Press.

Orend, Brian. 2006. *The Morality of War*. Toronto: Broadview Press.

Osiel, Mark. 2009. *The End of Reciprocity: Terror, Torture, and the Laws of War*. Cambridge: Cambridge University Press.

Owens, William A. and Edward Offley. 2000. *Lifting the Fog of War*. New York: Farrar, Straus & Giroux.

Page, Lewis. 2008. 'Snipers: Cowardly Assassins, or Surgical Soldiers?' Accessed 12 January 2015. http://www.theregister.co.uk/2008/11/28/sniper_feature?page=2.

Painter, Sidney. 1969. *French Chivalry: Chivalric Ideas and Practices in Mediaeval France*. Ithaca, NY: Cornell University Press.

Palmer, Ian. 2008. *Unmanned Aerial Vehicles: Robotic Air Warfare 1917–2007*. Oxford and New York: Osprey Publishing.

Panetta, Leon. 2009. 'Director's Remarks at the Pacific Council on International Policy'. Accessed 11 September 2017. https://www.cia.gov/news-information/speeches-testimony/directors-remarks-at-pacific-council.html.

Pardew, James W. 1991. 'The Iraqi Army's Defeat in Kuwait'. *Parameters* 21 (4): 17–23.

Paris, Michael. 1993. 'The Rise of the Airmen: The Origins of Air Force Elitism, c. 1890–1918'. *Journal of Contemporary History* 28 (1): 123–41.

Parker, Geoffrey. 1994. 'Early Modern Europe'. In *The Laws of War: Constraints on Warfare in the Western World*, edited by G. J. Andreopoulos, M. R. Shulman, and M. Howard. New Haven, CT: Yale University Press.

Parks, W. H. 1989. *Memorandum of Law: Executive Order 12333 and Assassination*. Department of the Army Pamphlet 27-50-204.

Parks, W. H. 2005. 'Conventional Weapons and Weapons Reviews'. *Yearbook of International Law* 8: 55–142.

Paskins, Barrie and Michael Dockrill. 1979. *The Ethics of War*. London: Duckworth.

Patterson, Orlando. 1982. *Slavery and Social Death: A Comparative Study*. Cambridge, MA: Harvard University Press.

Payne-Gallwey, Ralph. 1903. *The Book of the Crossbow*. London: Longmans, Green, and Co.

Pearson, Lee. 1969. 'Developing the Flying Bomb'. In *Naval Aviation in World War I*, edited by A. O. V. Wyen. Washington, DC: Chief of Naval Operations.

Pegler, Martin. 2006. *Out of Nowhere: A History of the Military Sniper*. Oxford: Osprey Publishing.

Pegler, Martin. 2008. *Sniping in the Great War*. Barnsley: Pen & Sword Military.

Perlez, Jane. 2013. 'Chinese Plan to Kill Drug Lord with Drone Highlights Military Advances'. Accessed 13 July 2017. http://www.nytimes.com/2013/02/21/world/asia/chinese-plan-to-use-drone-highlights-military-advances.html.

Peters, Ralph. 2009. 'The Rule Murdering Our Troops'. Accessed 11 June 2017. http://nypost.com/2009/09/24/the-rules-murdering-our-troops/.

Peterson, Richard. 2007. 'WWII Brutality Richard "Bud" Peterson P-51 Ace Interview'. Accessed 12 March 2015. https://www.youtube.com/watch?v=Q8LVlYJ5eJU.

Pevehouse, Jon C. and Joshua S. Goldstein. 1999. 'Serbian Compliance or Defiance in Kosovo'. *The Journal of Conflict Resolution* 43 (4): 538–46.

Pfaff, Tony. 2011. 'Risk, Military Ethics and Irregular Warfare'. Accessed 12 July 2017. http://www.fpri.org/article/2011/12/risk-military-ethics-and-irregular-warfare/.

Pilger, John. 1994. *Distant Voices*. London: Vintage.

Pilkington, Ed. 2015. 'Life As a Drone Operator: "Ever Step on Ants and Never Give it Another Thought?"' Accessed 11 December 2015. http://www.theguardian.com/world/2015/nov/18/life-as-a-drone-pilot-creech-air-force-base-nevada.

Pindell, Richard. 1993. 'The Most Dangerous Set of Men'. *Civil War Times Illustrated* (July–August).

Pizan, Christine de. 1999. *The Book of Deeds of Arms and of Chivalry*. Edited by C. C. Willard. Translated by S. Willard. University Park, PA: Pennsylvania State University Press.

Plaster, John. 2008. *The History of Sniping and Sharpshooting*. Boulder, CO: Paladin Press.

Plato. 1988. *The Laws of Plato*. Translated by T. L. Pangle. Chicago, IL: University of Chicago Press.

Plato. 2000. *The Republic*. Translated by T. Griffith. Cambridge: Cambridge University Press.

Plaw, Avery. 2010. 'Sudden Justice'. 7th Annual Global Conference on War and Peace, Prague 1.

Polybius. 2003. 'The Histories of Polybius, Fragments of Book XIII'. Accessed 17 November 2014. http://penelope.uchicago.edu/Thayer/E/Roman/Texts/Polybius/13*.html.

Ponte, Carla Del. 2000. 'United Nations International Criminal Tribunal for the Former Yugoslavia on the investigation and prosecution of crimes committed in Kosovo'.

Accessed 4 July 2013. http://www.icty.org/en/press/final-report-prosecutor-committee-established-review-nato-bombing-campaign-against-federal.

Porat, Iddo and Bohrer Ziv. 2014. 'Preferring One's Own Civilians: May Soldiers Endanger Enemy Civilians More Than They Would Endanger Their State's Civilians?' *The George Washington International Law Review* 47 (1): 99–158.

Power, Matthew. 2013. 'Confessions of a Drone Warrior'. Accessed 10 November 2015. http://www.gq.com/story/drone-uav-pilot-assassination?currentPage=3.

Power, Tracey J. 1998. *Lee's Miserables: Life in the Army of Northern Virginia from the Wilderness to Appomattow*. Chapel Hill, NC: University of North Carolina Press.

Price, Richard M. 2007. *The Chemical Weapons Taboo*. Ithaca, NY: Cornell University Press.

Pritchett, William K. 1974. *The Greek State at War: Part II*. Berkeley, CA: University of California Press.

Pryer, Douglas A. 2013. 'The Rise of the Machines: Why Increasingly "Perfect" Weapons Help Perpetuate Our Wars and Endanger Our Nation'. *Military Review* 93 (2): 14–23.

Pufendorf, Samuel von. 1729. *On the Law of Nature and Nations: Eight Books*. Translated by B. Kennett. London: Printed for J. Walthoe, R. Wilkin, J. and J. Bonwicke, S. Birt, T. Ward, and T. Osborne. Accessed 20 July 2019. https://ia800202.us.archive.org/19/items/oflawofnaturenat00pufe/oflawofnaturenat00pufe.pdf.

Purdon, James. 2010. 'Bomber Country by Daniel Swift: An Attempt to Reclaim the Forgotten Work and Lives of Airborne War Poets is Admirable but Doesn't Always Hit the Target'. Accessed 29 August 2017. https://www.theguardian.com/books/2010/aug/29/bomber-county-daniel-swift-review.

Purkiss, Jessica. 2019. 'Trump Rolls Back Transparency on Civilian Drone Deaths'. *The Bureau of Investigative Journalism*. Accessed 10 March 2019. https://www.thebureauinvestigates.com/stories/2019-03-07/trump-tells-cia-dont-tell-world-about-dead-civilians.

Raleigh, Sir Walter. 1922. *The War in the Air*. Vol. 1. Oxford: Clarendon Press.

Ralph, Jason. 2013. *America's War on Terrorism: The State of the 9/11 Exception from Bush to Obama*. Oxford: Oxford University Press.

Ramsey, Paul. 1968. *The Just War: Force and Political Responsibility*. New York: Charles Scribner's Sons.

Rattenbury, Richard C. 2015. 'Rifle Weapon'. Accessed 23 February 2015. http://www.britannica.com/EBchecked/topic/503347/rifle.

Rawlings, Louis. 2007. *The Ancient Greeks at War*. Manchester: Manchester University Press.

Rawls, John. 1955. 'Two Concepts of Rules'. *The Philosophical Review* 64 (1): 3–32.

Rawls, John. 1999. 'Fifty Years After Hiroshima'. In Rawls, *Collected Papers*. Edited by S. Freeman. Cambridge, MA: Harvard University Press.

Ray, Fred L. 2006. 'The Killing of Uncle John'. *Civil War Times* 45 (4): 26–32.

Record, Jeffrey. 2000. 'Force-Protection Fetishism'. *Aerospace Power Journal* 14 (2): 4–11.

Reichberg, Gregory M. 2008. 'Just War and Regular War: Competing Paradigms'. In *Just and Unjust Warriors: The Moral and Legal Status of Soldiers*, edited by D. Rodin and H. Shue. Oxford: Oxford University Press.

Reichberg, Gregory M. 2013. 'Culpability and Punishment in Classical Theories of Just War'. In *Just War: Authority, Tradition, and Practice*, edited by C. O'Driscoll, J. Williams, and A. F. Lang Jr. Washington, DC: Georgetown University Press.

Reichberg, Gregory M., Henrik Syse, and Endre Begby, eds. 2006. *The Ethics of War: Classical and Contemporary Readings*. Malden, MA: Blackwell Publishing.

Reid, Brian Holden. 2002. 'The Influence of the Vietnam Syndrome on the Writing of Civil War History'. *The RUSI Journal* 147 (1): 44–52.

Reinhard, Marcel R. 1950. *Le Grand Carnot*. Vol. 2. Paris: Hachette.

Remarque, Erich M. 1929. *All Quiet on the Western Front*. New York: Fawcett Crest.

Reppenhagen, Garett. 2015. 'I Was An American Sniper, and Chris Kyle's War Was Not My War'. Accessed 13 June 2015. http://www.salon.com/2015/02/01/i_was_an_american_sniper_and_chris_kyle%E2%80%99s_war_was_not_my_war/.

Reus-Smit, Christian. 2008. 'Reading History Through Constructivist Eyes'. *Millennium: Journal of International Studies* 37 (2): 395–414.

Reuters. 2018. 'U.S. Confirms Deployment of Armed Drones in Niger'. Accessed 11 January 2019. https://www.reuters.com/article/us-usa-niger-security-drones/u-s-confirms-deployment-of-armed-drones-in-niger-idUSKBN1KK18C.

Rhetorica ad Herennium. 1954. Accessed 11 July 2017. http://penelope.uchicago.edu/Thayer/E/Roman/Texts/Rhetorica_ad_Herennium/4B*.html.

Rhodes, Richard. 2005. 'Nuclear Options'. Accessed 26 January 2017. http://www.nytimes.com/2005/05/15/books/review/nuclear-options.html.

Riza, Shane M. 2013. *Killing Without Heart: Limits on Robotic Warfare in an Age of Persistent Conflict*. Washington, DC: Potomac Books.

Rizzo, John Anthony. 2014. *Company Man: Thirty Years of Controversy and Crisis in the CIA*. New York: Scribner.

Roberts, Adam. 1999. 'NATO's "Humanitarian War" Over Kosovo'. *Survival* 41 (3): 102–23.

Roberts, Adam. 2008. 'The Principle of Equal Application of the Laws of War'. In *Just and Unjust Warriors: The Moral and Legal Status of Soldiers*, edited by D. Rodin and H. Shue. Oxford: Oxford University Press.

Robertson, James I. 1988. *Soldiers, Blue and Grey*. Columbia, SC: University of South Carolina Press.

Robinson, Paul. 1999. '"Ready to Kill but Not to Die": NATO Strategy in Kosovo'. *International Journal* 54 (4): 671–82.

Robinson, Paul. 2006. *Military Honour and the Conduct of War: From Ancient Greece to Iraq*. Abingdon: Routledge.

Robinson, Paul. 2008. 'Introduction: Ethics Education in the Military'. In *Ethics Education in the Military*, edited by P. Robinson, N. de Lee, and D. Carrick. Aldershot: Ashgate.

Rodin, David. 2002. *War and Self-Defence*. Oxford: Clarendon Press.

Rodin, David. 2006. 'The Ethics of Asymmetric War'. In *The Ethics of War: Shared Problems in Different Traditions*, edited by R. Sorabji and D. Rodin. Aldershot: Ashgate.

Rodin, David. 2008. 'The Moral Inequality of Soldiers: Why Jus in Bello Asymmetry is Half Right'. In *Just and Unjust Warriors: The Moral and Legal Status of Soldiers*, edited by D. Rodin and H. Shue. Oxford: Oxford University Press.

Rodin, David and Henry Shue, eds. 2008. *Just and Unjust Warriors: The Moral and Legal Status of Soldiers*. Oxford: Oxford University Press.

Rogers, A. P. V. 2000. 'Zero-Casualty Warfare'. *International Review of the Red Cross* 82 (837): 165–82.

Rohde, David. 2012. 'The Drone War'. Accessed 11 October 2015. http://www.reuters.com/article/2012/01/17/davos-reutersmagazine-dronewar-idAFL1E8CHCXX20120117.

Rohde, David and Mohammed Khan. 2004. 'The Reach of War: Militants; Ex-Fighter for Taliban Dies in Strike in Pakistan'. Accessed 12 September 2016. http://www.nytimes.com/2004/06/19/world/the-reach-of-war-militants-ex-fighter-for-taliban-dies-in-strike-in-pakistan.html.

Roth, Kenneth. 2017. 'Must it Always be Wartime?' Accessed 5 March 2017. http://www.nybooks.com/articles/2017/03/09/must-it-always-be-wartime/.

Rothenberg, Gunther. 1994. 'The Age of Napoleon'. In *The Laws of War: Constraints on Warfare in the Western World*, edited by G. J. Andreopoulos, M. R. Shulman, and M. Howard. New Haven, CT: Yale University Press.

Rothenberg, Gunther. 1997. *The Art of Warfare in the Age of Napoleon*. Staplehurst: Spellmount.

Rousseau, Jean-Jacques. 1968. *The Social Contract*. Vol. 1. Translated by M. Cranston. Harmondsworth: Penguin.

Rousseau, Jean-Jacques. 1992. *Discourse on the Science and Arts: (First Discourse) and Polemics*. Translated by J. R. Bush, R. D. Masters, C. Kelly, and T. Marshall. Hanover, NH: University Press of New England.

Rousseau, Jean-Jacques. 1997. 'Du Contrat Social, ou Principes du Droit Politique'. In *Rousseau: The Social Contract and Other Later Political Writings*. Edited and Translated by V. Gourevitch. Cambridge: Cambridge University Press.

Royal Air Force. 2008. *Ethos, Core Values and Standards*. Air Publication 1, 2nd edition.

Rumsfeld, Donald. 2011. *Known and Unknown: A Memoir*. New York: Sentinel.

Ryan, Cheney. 2008. 'Moral Equality, Victimhood, and the Sovereignty Symmetry Problem'. In *Just and Unjust Warriors: The Moral and Legal Status of Soldiers*, edited by D. Rodin and H. Shue. Oxford: Oxford University Press.

Ryan, Klem. 2012. 'Fair and Unfair Wars'. D.Phil thesis, Department of Politics and International Relations, University of Oxford.

Ryan, Klem. 2014. 'What's Wrong with Drones? The Battlefield in International Humanitarian Law'. In *The American Way of Bombing: Changing Ethical and Legal Norms, from Flying Fortresses to Drones*, edited by M. Evangelista and H. Shue. Ithaca, NY: Cornell University Press.

Sabel, Robbie. 2000. 'Chivalry in the Air?' In *International Law Across the Spectrum of Conflict: Essays in Honour of Professor L. C. Green on the Occasion of his Eightieth Birthday*, edited by M. N. Schmitt. Newport, RI: Naval War College.

Sadowski, Yahya M. 1998. *The Myth of Global Chaos*. Washington, DC: Brookings Institution Press.

Sanchez, Ricardo S. and Donald T. Phillips. 2008. *Wiser in Battle: A Soldier's Story*. New York: HarperCollins.

Sandoz, Yves, Christophe Swinarski, and Bruno Zimmermann. 1987. *Commentary on the Additional Protocols of 8 June 1977 to the Geneva Conventions of 12 August 1949*. Geneva: Martin Nijhoff.

Sassòli, Marco and Laura M. Olson. 2008. 'The Relationship Between International Humanitarian and Human Rights Law Where it Matters: Admissible Killing and Internment of Fighters in Non-International Armed Conflicts'. *International Review of the Red Cross* 90 (871): 599–627.

Sauer, Frank and Niklas Schörnig. 2012. 'Killer Drones: "The Silver Bullet" of Democratic Warfare?' *Security Dialogue* 43: 363–80.

Savage, Charlie and Eric Schmitt. 2017. 'Trump Poised to Drop Some Limits on Drone Strikes and Commando Raids'. Accessed 24 September 2017. https://www.nytimes.com/2017/09/21/us/politics/trump-drone-strikes-commando-raids-rules.html.

Scahill, Jeremy. 2015. 'The Assassination Complex'. Accessed 10 November 2016. https://theintercept.com/drone-papers/the-assassination-complex/.

Scahill, Jeremy and Glenn Greenwald. 2014. 'The NSA's Secret Role in the U.S. Assassination Program'. Accessed 13 February 2019. https://theintercept.com/2014/02/10/the-nsas-secret-role/.

Scales, Robert H. 2005. 'Urban Warfare: A Soldier's View'. *Military Review* 85 (1): 9–18.

Scarry, Elaine. 1985. *The Body in Pain: The Making and Unmaking of the World*. New York and Oxford: Oxford University Press.

Scheipers, Sibylle. 2011. 'The Status and Protections of Prisoners of War and Detainees'. In *The Changing Character of War*, edited by H. Strachan and S. Scheipers. New York and Oxford: Oxford University Press.

Schell, Jonathan. 2011. 'Attacking Libya – and the Dictionary'. *Los Angeles Times*, 21 June.

Schelling, Thomas C. 1980. *The Strategy of Conflict*. Cambridge, MA: Harvard University Press.

Schmitt, Carl. 1999. 'Die Wendung zum Diskriminierenden Kriegsbegriff'. In *The Challenge of Carl Schmitt*, edited by C. Mouffe. London: Verso.

Schmitt, Carl. 2006. *The Nomos of the Earth in the International Law of the Jus Publicum Europaeum*. Translated by G. L. Ulmen. New York: Telos Press Publishing.

Schmitt, Carl. 2007. *Theory of the Partisan: Intermediate Commentary on the Concept of the* Political. New York: Telos Press.

Schmitt, Eric. 1999. 'Crisis in the Balkans: Military Analysis; Price of Civilian Deaths: NATO Debit, Serb Asset'. Accessed 10 May 2017. http://www.nytimes.com/1999/04/15/world/crisis-balkans-military-analysis-price-civilian-deaths-nato-debit-serb-asset.html.

Schmitt, Michael N. 2006. 'War, Technology and the Law of Armed Conflict'. In *The Law of War in the 21st Century: Weaponry and the Use of Force*. Vol. 82, edited by A. M. Helm. Newport, RI: Naval War College.

Schmitt, Michael N. 2007. '21st Century Conflict: Can the Law Survive?' *Melbourne Journal of International Law* 8 (2): 443–76.

Schulzinger, Robert D. 2006. *A Time for Peace: The Legacy of the Vietnam War*. New York: Oxford University Press.

Schulzke, Marcus. 2016. 'Rethinking Military Virtue Ethics in an Age of Unmanned Weapons'. *Journal of Military Ethics* 15 (3): 187–204.

Schwarzkopf, H. Norman. 1992. *It Doesn't Take a Hero: The Autobiography of General H. Norman Schwarzkopf*. London: Bantam.

Science for the People. 1973. 'Toys Against the People'. *Bi-Monthly Publication of Scientists and Engineers for Social and Political Action* 5 (1): 8–10, 37–41.

Sciolino, Elaine. 1998. 'The World: Theatre of War; The New Face of Battle Wears Greasepaint'. Accessed 11 April 2017. http://www.nytimes.com/1998/02/22/weekinreview/the-world-theater-of-war-the-new-face-of-battle-wears-greasepaint.html.

Searle, John R. 1995. *The Construction of Social Reality*. London: Penguin Books.

Second Lateran Council—1139 A.D. 1139. Accessed 11 July 2016. http://www.papalencyclicals.net/councils/ecum10.htm.

Selesky, Harold E. 1994. 'Colonial America'. In *The Laws of War: Constraints on Warfare in the Western World*, edited by G. J. Andreopoulos, M. R. Shulman, and M. Howard. New Haven, CT: Yale University Press.

Shane, Scott. 2011. 'Coming Soon: The Drone Arms Race'. Accessed 11 November 2016. http://www.nytimes.com/2011/10/09/sunday-review/coming-soon-the-drone-arms-race.html.

Shane, Scott. 2016. 'Drone Strike Statistics Answer Few Questions, and Raise Many'. Accessed 13 October 2017. https://www.nytimes.com/2016/07/04/world/middleeast/drone-strike-statistics-answer-few-questions-and-raise-many.html.

Shanker, Thom. 2013. 'A New Medal Honors Drone Pilots and Computer Experts'. *The New York Times*, 14 February.

Shapiro, Scott J. 2007. 'The "Hart–Dworkin" Debate: A Short Guide for the Perplexed'. *Public Law and Legal Theory Working Paper Series* 77: 1–54.

Sharkey, Noel. 2008. 'Cassandra or False Prophet of Doom: AI Robots and War'. *Intelligent Systems, IEEE* 23 (4): 14–17.

Sharkey, Noel. 2010. 'Saying "No!" to Lethal Autonomous Targeting'. *Journal of Military Ethics* 9 (4): 369–83.

Shaw, Martin. 2005. *The New Western Way of War*. Cambridge: Polity Press.

Shay, Jonathan. 2003. *Achilles in Vietnam: Combat Trauma and the Undoing of Character*. New York: Scribner.

Sherman, William C. 1919. 'Tentative Manual for the Employment of Air Service'. *The U.S. Air Service* 2: 313–408.

Sherman, William T. 1864. 'William T. Sherman to Henry W. Halleck, December 24, 1864'. Accessed 20 February 2015. https://cwnc.omeka.chass.ncsu.edu/items/show/144.

Sherry, Michael. 1987. *The Rise of American Air Power: The Creation of Armageddon*. New Haven, CT: Yale University Press.

Short, Michael C. 1999. 'War in Europe – Interview'. Accessed 14 June 2013. http://www.pbs.org/wgbh/pages/frontline/shows/kosovo/interviews/short.html.

Shue, Henry. 2003. 'Bombing to Rescue? NATO's 1999 Bombing of Serbia'. In *Ethics and Foreign Intervention*, edited by D. K. Chatterjee and D. E. Scheid. Cambridge: Cambridge University Press.

Shue, Henry. 2004. 'Torture'. In *Torture: A Collection*, edited by S. Levinson. Oxford: Oxford University Press.

Shue, Henry. 2008. 'Do We Need a Morality of War?' In *Just and Unjust Warriors: The Moral and Legal Status of Soldiers*, edited by D. Rodin and H. Shue. Oxford: Oxford University Press.

Shue, Henry. 2011. 'Civilian Protection and Force Protection'. In *Ethics, Law and Strategy*, edited by D. Whetham. Basingstoke: Palgrave Macmillan.

Shue, Henry. 2013. 'Laws of War, Morality, and International Politics: Compliance, Stringency, and Limits'. *Leiden Journal of International Law* 26 (2): 271–92.

Shue, Henry. 2016. 'Target-Selection Norms, Torture Norms, and Growing US Permissiveness'. In Shue, *Fighting Hurt: Rule and Exception in Torture and War*. Oxford: Oxford University Press.

Shue, Henry and David Wippman. 2002. 'Limiting Attacks on Dual-Use Facilities Performing Indispensable Civilian Functions'. *Cornell International Law Journal* 35 (3): 559–79.

Sidebottom, Harry. 2004. *Ancient Warfare: A Very Short Introduction*. Oxford: Oxford University Press.

Sifton, John. 2012. 'A Brief History of Drones'. Accessed 10 October 2015. http://www.thenation.com/article/brief-history-drones/.

Simmons, George. 1899. *A British Rifle Man: The Journals and Correspondence of Major George Simmons, Rifle Brigade, During the Peninsula War and the Campaign of Waterloo*. Edited by W. Verner. London: A. & C. Black.

Singer, Peter W. 2009. 'Robots at War: The New Battlefield'. *The Wilson Quarterly* 33 (1): 30–48.

Singer, Peter W. 2010. 'Interview with Defence Expert P. W. Singer: "The Soldiers Call it War Porn"'. Accessed 9 November 2015. http://www.spiegel.de/international/world/interview-with-defense-expert-p-w-singer-the-soldiers-call-it-war-porn-a-682852.html.

Smith, Adam. 1978. *Lectures on Jurisprudence*. Oxford: Clarendon Press.

Smith, David Livingstone. 2011. *Less than Human: Why We Demean, Enslave, and Exterminate Others*. New York: St. Martin's Press.

Smith, Malcolm. 1984. *British Air Strategy Between the Wars*. Oxford: Clarendon Press.
Smith, Rupert. 2005. *The Utility of Force: The Art of War in the Modern World*. London: Allen Lane.
Smith, Thomas W. 2002. 'The New Law of War: Legitimising Hi-Tech and Infrastructural Violence'. *International Studies Quarterly* 46: 355–74.
Smythe, John. 1590. *Certen Discourses Concerning the Formes and Effects of Diuers Sorts of Weapons*. Accessed 12 July 2019. https://archive.org/details/certaindiscourse00smyt
Solis, Gary D. 2010. *The Law of Armed Conflict: International Humanitarian Law in War*. Cambridge and New York: Cambridge University Press.
Spaight, J. M. 1911. *War Rights on Land*. London: Macmillan and Co.
Spaight, J. M. 1930. *Air Power and Cities*. New York: Longmans.
Spaight, J. M. 1938. 'The Chaotic State of the International Law Governing Bombardment'. *Royal Air Force Quarterly* 9 (1): 24–32.
Sparrow, Robert. 2012. '"Just Say No" to Drones'. *Technology and Society Magazine, IEEE* 31 (1): 56–63.
Sparrow, Robert. 2013. 'War Without Virtue?' In *Killing by Remote Control: The Ethics of an Unmanned Military*, edited by B. J. Strawser. Oxford: Oxford University Press.
Spetalnick, Matt. 2014. 'U.S. Drone Strikes Risk "Slippery Slope" to Endless War, Panel Says'. Accessed 29 June 2014. http://www.reuters.com/article/2014/06/26/us-usa-security-drones-idUSKBN0F12S920140626.
Spiegel, Paul B. and Peter Salama. 2000. 'War and Mortality in Kosovo. 1998–99: An Epidemiological Testimony'. *The Lancet* 355: 2204–9.
Spinetta, Lawrence. 2010. 'Secrets Behind the Surprising Rise of Drones. Accessed 11 September 2016. http://www.historynet.com/the-rise-of-unmanned-aircraft.htm.
Stacey, Robert C. 1994. 'The Age of Chivalry'. In *The Laws of War: Constraints on Warfare in the Western World*, edited by G. J. Andreopoulos, M. R. Shulman, and M. Howard. New Haven, CT: Yale University Press.
Stafford, David. 2007. *Endgame, 1945: The Missing Final Chapter of World War II*. New York: Little, Brown and Company.
Stanford Law School—International Human Rights and Conflict Resolution Clinic & NYU School of Law Global Justice Clinic. 2012. 'Living under Drones: Death, Injury and Trauma to Civilians from US Drone Practices in Pakistan'. Accessed 23 May 2017. https://chrgj.org/wp-content/uploads/2016/09/Living-Under-Drones.pdf.
Statman, Daniel. 2011. 'Can Wars Be Fought Justly? The Necessity Condition Put to the Test'. *Journal of Moral Philosophy* 8 (3): 435–51.
Statman, Daniel. 2012. 'Can Just War Theory Justify Targeted Killing? Three Possible Models'. In *Targeted Killings: Law and Morality in an Asymmetrical World*, edited by J. D. Ohlin, A. Altman, and C. Finkelstein. Oxford: Oxford University Press.
Steevens, G. W. 1898. *With Kitchener to Khartum*. New York: Dodd, Mead.
Steinbeck, John. 1942. *Bombs Away: The Story of a Bomber Team*. New York: Penguin.
Steinhoff, Uwe. 2013. 'Killing them Safely: Extreme Asymmetry and its Discontents'. In *Killing in Remote Control*, edited by B. J. Strawser. Oxford: Oxford University Press.
Stirling, Robert. 2012. *Special Forces Sniper Skills*. Oxford: Osprey Publishing.
Strachan, Hew. 2010. 'The Changing Character of War in a Multi-Polar World'. *New Zealand International Review* 35 (1): 2–7.
Strachan, Hew. 2014. 'Heroic Warfare and the Problem of Mass Armies: France 1871–1914'. In *Heroism and the Changing Character of War: Towards Post-Heroic Warfare?*, edited by S. Scheipers. Basingstoke: Palgrave Macmillan.

Strachan, Hew and Sibylle Scheipers. 2011. 'Introduction: The Changing Character of War'. In *The Changing Character of War*, edited by H. Strachan and S. Scheipers. Oxford: Oxford University Press.

Strawser, Bradley. 2010. 'Moral Predators: The Duty to Employ Uninhabited Aerial Vehicles'. *Journal of Military Ethics* 9 (4): 342–68.

Strawser, Bradley. 2011. 'Walking the Tightrope of Just War'. *Analysis* 71 (3): 533–44.

Suárez, Francisco. 2006. 'Disputation XIII. Section VII: What is the Proper Mode (Debitus Modus) of Conducting War?' In *The Ethics of War: Classical and Contemporary Readings*, edited by H. Syse, E. Begby, and G. M. Reichberg. Malden, MA: Blackwell Publishing.

Suskind, Ron. 2007. *The One Percent Doctrine: Deep Inside America's Pursuit of its Enemies Since 9/11*. London: Pocket Books.

Tacitus. 2016. *Germania*. Oxford: Oxford University Press.

Tägil, Sven. 2014. 'Alfred Nobel's Thoughts about War and Peace'. Accessed 20 January 2017. https://www.nobelprize.org/alfred-nobel/alfred-nobels-thoughts-about-war-and-peace/.

Talley, Sharon. 2009. *Ambrose Bierce and the Dance of Death*. Knoxville, TN: University of Tennessee Press.

Taylor, Claire. 2011. 'Armed Forces Covenant'. Accessed 2 October 2013. https://assets.publishing.service.gov.uk/government/uploads/system/uploads/attachment_data/file/395359/SN05979_Military_Covenant.pdf

Taylor, Jared. 2011. 'The Cultural Impact of the American Sniper Program'. *Utah Historical Review* 1: 87–102.

Taylor, Telford. 1970. *Nuremberg and Vietnam: An American Tragedy*. Chicago, IL: Quadrangle Books.

Telhami, Shibley. 2004. *The Stakes: America in the Middle East – The Consequences of Power and the Choice of Peace*. Boulder, CO: Westview Press.

Tenfold, Vegas. 2016. 'The Untold Casualties of the Drone War'. Accessed 5 March 2016. http://www.rollingstone.com/politics/news/the-untold-casualties-of-the-drone-war-20160218.

Tennyson, Lord Alfred. 1935. *Works: With Notes by the Author*. New York: Macmillan.

Terry, J. D. 1975. *The Evolving Law of Aerial Warfare*. Montgomery, AL: U.S. Air University, Air War College.

Tesla, Nikola. 1898. 'Plans to Dispense with Artillery of the Present Type'. *The Sun*, 21 November. Accessed 12 July 2019. http://www.tfcbooks.com/tesla/1898-11-21.htm.

Tesón, Fernando R. 2012. 'Targeted Killing in War and Peace: A Philosophical Analysis'. In *Targeted Killings: Law and Morality in an Asymmetrical World*, edited by J. D. Ohlin, A. Altman, and C. Finkelstein. Oxford: Oxford University Press.

The Brussels Declaration. 1874. 'Project of an International Declaration concerning the Laws and Customs of War (Brussels, 27 August 1874)'. Accessed 20 March 2015. http://web.ics.purdue.edu/~wggray/Teaching/His300/Handouts/Brussels-1874.html.

The Bureau of Investigative Journalism. 2016. 'Drone Wars: The Full Data'. Accessed 10 November 2016. https://www.thebureauinvestigates.com/category/projects/drones/drones-graphs/.

The Convention on Certain Conventional Weapons. 1995. 'Protocol on Blinding Laser Weapons'. Accessed 16 November 2016. http://www.unog.ch/80256EDD006B8954/(httpAssets)/8463F2782F711A13C12571DE005BCF1A/$file/PROTOCOL+IV.pdf.

The Economist. 2012. 'Death From Afar'. Accessed 13 May 2017. http://www.economist.com/news/international/21565614-america-uses-drones-lot-secret-and-largely-unencumbered-declared-rules-worries.

The Guardian. 2016. 'ISIS Booby-Trapped Drone Kills Troops in Iraq, Officials Say'. Accessed 10 November 2016. https://www.theguardian.com/world/2016/oct/12/exploding-drone-sent-by-isis-allies-kills-and-wounds-troops-in-iraq-report

The New York Times. 1880. 'The Hessians at the Battle of Long Island'. *The New York Times*, 27 December. Accessed 14 July 2019. https://www.nytimes.com/1880/12/27/archives/britains-hired-troops-this-hessians-at-the-battle-of-long-island.html

The Spectator. 1913. 'The Air, the Public, and the Future'. Accessed 13 July 2015. http://archive.spectator.co.uk/article/17th-may-1913/9/the-air-the-public-and-the-future.

The Western Front Association. 2008. 'The British "School of Sniping" On the Western Front'. Accessed 12 April 2015. http://westfrontassoc.mtcdevserver.com/the-great-war/great-war-on-land/weapons-equipment-uniform/118-school-sniping.html#sthash.cO4gVYJG.dpbs.

Thomas, Evan. 2009. 'General McChrystal's Plan for Afghanistan'. Accessed 12 July 2017. http://www.newsweek.com/general-mcchrystals-plan-afghanistan-79551.

Thomas, Ward. 2001. *The Ethics of Destruction: Norms and Force in International Relations*. Ithaca, NY: Cornell University Press.

Thomson, Judith J. 1991. 'Self-Defense'. *Philosophy & Public Affairs* 20 (4): 283–310.

Thornton, Thomas. 1806. *A Sporting Tour Through Various Parts of France in the Year 1802*. Vol. 2. London: Longman, Hurst, Rees and Orme, and C. Chapple.

Tilford, Earl H. Jr. 1999. 'Operation Allied Force and the Role of Air Power'. *Parameters* 29 (4): 24–38.

Toffler, Alvin and Heidi Toffler. 1993. *War and Anti-War: Survival at the Dawn of the Twenty-First Century*. Boston: Little, Brown.

Trask, David. 1999. 'Military Imagination in the United States, 1815–1917'. In *Anticipating Total War*, edited by R. Chikering, S. Forster, and M. F. Boemeke. Cambridge: Cambridge University Press.

Truver, Scott C. 2000. 'Operation Allied Force: The Lessons Learned'. *Sea Power* 23 (6): 35–7.

Turns, David. 2014. 'Droning On: Some International Humanitarian Law Aspects of the Use of Unmanned Aerial Vehicles in Contemporary Armed Conflicts'. In *Contemporary Challenges to the Laws of War: Essays*, edited by J. Summers, N. D. White, and C. Harvey. Cambridge: Cambridge University Press.

Turse, Nick, Henrik Moltke, and Alice Speri. 2018. 'Secret War: The U.S. Has Conducted 550 Drone Strikes in Libya Since 2011 – More Than in Somalia, Yemen, or Pakistan'. Accessed 20 August 2019. https://theintercept.com/2018/06/20/libya-us-drone-strikes/.

United Kingdom Ministry of Defence. 2004. *The Manual of the Law of Armed Conflict*. Oxford: Oxford University Press.

United Kingdom Ministry of Defence. 2011. 'Joint Doctrine Note 2/11 The UK Approach to Unmanned Aircraft Systems'. Accessed 20 March 2015. https://www.gov.uk/government/uploads/system/uploads/attachment_data/file/33711/20110505JDN_211_UAS_v2U.pdf.

United States Army. 1914. 'FM 27-10, Law of Land Warfare Basic Field Manual'. Accessed 12 November 2014. http://www.loc.gov/rr/frd/Military_Law/pdf/rules_warfare-1914.pdf.

United States Army. 1940. 'FM 27-10, Law of Land Warfare Basic Field Manual'. Accessed 3 September 2015. https://ia701202.us.archive.org/1/items/Fm27-10/Fm27-10.pdf.

United States Army. 1956. 'FM 27-10, Law of Land Warfare Basic Field Manual'. Accessed 13 November 2014. https://www.loc.gov/rr/frd/Military_Law/pdf/law_warfare-1956.pdf.

United States Army. 1984. 'FM 27-2, Your Conduct in Combat Under the Law of War'. Accessed 10 May 2016. https://www.loc.gov/rr/frd/Military_Law/pdf/conduct-in-combat-1984.pdf.

United States Army. 1994. 'FM 23-10, Sniper Training'. Accessed 10 April 2015. https://archive.org/details/milmanual-fm-23-10-sniper-training.

United States Army. 2000. 'FM 27-100, Legal Support to Operations'. Accessed 3 March 2016. https://www.loc.gov/rr/frd/Military_Law/pdf/legal_support_operations.pdf.

United States Army. 2002. 'FM 3-06.11, Combined Arms Operations in Urban Terrain'. Accessed 11 May 2017. http://www.globalsecurity.org/military/library/policy/army/fm/3-06-11/ch6.htm.

United States Army. 2006. 'FM 3-24, Counterinsurgency'. Accessed 11 December 2015. https://www.hsdl.org/?abstract&did=468442.

United States Department of Defense. 2015. 'Law of War Manual'. Accessed 13 March 2016. https://archive.defense.gov/pubs/law-of-war-manual-june-2015.pdf.

United States Department of Justice. 2013. 'Lawfulness of a Lethal Operation Directed Against a U.S. Citizen Who is a Senior Operational Leader of Al-Qa'ida or An Associated Force'. Accessed 23 May 2017. https://fas.org/irp/eprint/doj-lethal.pdf.

van Creveld, Martin. 1989. *Technology and War: From 2000 B.C. to the Present*. New York: Free Press.

van Creveld, Martin. 1991. *The Transformation of War*. New York: Free Press.

van Creveld, Martin. 2010. 'The Rise and Fall of Air Power'. In *A History of Air Warfare*, edited by J. A. Olsen. Washington, DC: Potomac Books.

van Creveld, Martin. 2011. *The Age of Airpower*. New York: Public Affairs.

Van Damme, Guy. 2002. 'A Historical Overview of the Jus in Bello Constraints'. In *Moral Constraints on War: Principles and Cases*, edited by B. Coppieters and N. Fotion. Lanham, MD: Lexington Books.

Van Evera, Stephen. 1984. 'The Cult of the Offensive and the Origins of the First World War'. *International Security* 9 (1): 58–107.

Vanderbilt, Tom. 2010. *Survival City: Adventures Among the Ruins of Atomic America*. Chicago, IL: University of Chicago Press.

Vattel, Emer de. 2006. 'War in Due Form'. In *The Ethics of War*, edited by H. Syse, E. Begby, and G. M. Reichberg. Oxford: Blackwell.

Vattel, Emer de. 2008. *The Law of Nations, or, Principles of the Law of Nature, Applied to the Conduct and Affairs of Nations and Sovereigns, With Three Early Essays on the Origin and Nature of Natural Law and on Luxury*. Edited by B. Kapossy and R. Whatmore. Translated by T. Nugent. Indianapolis, IN: Liberty Fund.

Verne, Jules. 1891. *The Clipper in the Clouds*. London: Sampson Low, Marston.

Vitoria, Francisco de. 1991. 'On the Law of War'. In *Vitoria: Political Writings*, edited by A. Pagden and J. Lawrence. Cambridge: Cambridge University Press.

Walker, Paul and Terence Lovat. 2016. 'Should We Be Talking About Ethics or About Morals?' *Ethics and Behavior* 27 (5): 436–44.

Walker, Samuel G. 2012. 'Lawful Murder: Unnecessary Killing in the Law of War'. *Canadian Journal of Law and Jurisprudence* 25 (2): 417–46.

Waller, Douglas and John Barry. 1992. 'The Day We Stopped the War'. *Newsweek*, 19 January.

Walzer, Michael. 1992. 'Justice and Injustice in the Gulf War'. In *But Was it Just? Reflections on the Morality of the Persian Gulf War*, edited by D. E. Decosse. New York: Doubleday.

Walzer, Michael. 2002. 'The Triumph of Just War Theory (And the Dangers of Success)'. *Social Research* 69 (4): 925–44.

Walzer, Michael. 2004a. 'Kosovo'. In *Arguing about War*, edited by M. Walzer. New Haven, CT: Yale University Press.

Walzer, Michael. 2004b. 'Two Kinds of Military Responsibility'. In *Arguing about War*, edited by M. Walzer. New Haven, CT: Yale University Press.

Walzer, Michael. 2004c. 'Terrorism: A Critique of Excuses'. In *Arguing about War*, edited by M. Walzer. New Haven, Ct: Yale University Press.

Walzer, Michael. 2006a. *Just and Unjust Wars: A Moral Argument with Historical Illustrations*. New York: Basic Books.

Walzer, Michael. 2006b. 'Terrorism and Just War'. *Philosophia* 34: 3–12.

Walzer, Michael. 2006c. 'Response to McMahan's Paper'. *Philosophia* 34 (1): 43–5.

Walzer, Michael. 2013a. 'Targeted Killing and Drone Warfare'. Accessed 21 November 2016. https://www.dissentmagazine.org/online_articles/targeted-killing-and-drone-warfare.

Walzer, Michael. 2013b. 'Coda: Can the Good Guys Win?' *The European Journal of International Law* 24 (1): 433–44.

Walzer, Michael and Avishai Margalit. 2009. 'Israel: Civilians and Combatants'. *The New York Review of Books*. Accessed 20 January 2019. https://www.nybooks.com/articles/2009/05/14/israel-civilians-combatants/

Ward, Alex. 2017. 'Guess Who Has Drones Now? ISIS'. Accessed 3 June 2017. https://www.vox.com/world/2017/5/30/15686240/drones-isis-iraq-syria.

Washington, George. 1932. *The Writings of George Washington*. Vol. 7. Washington, DC: U.S. Government Printing Office.

Webb, Tiger. 2015. 'What if Top Gun Was About Drones?' Accessed 20 March 2015. http://www.abc.net.au/radionational/programs/futuretense/what-if-top-gun-was-about-drones/6140152.

Weber, Max. 1946. *From Max Weber: Essays in Sociology*. Edited by H. H. Garth and C. W. Mills. London: Routledge & Kegan Paul.

Weber, Max. 1958. *The Protestant Ethic and the Spirit of Capitalism*. Translated by T. Parsons. New York: Charles Scribner's Sons.

Wells, H. G. 1908. *The War in the Air*. London: George Bell & Sons.

Welsh, Jennifer. 2015. 'The Morality of "Drone Warfare"'. In *Drones and the Future of Conflict: Ethical, Legal, and Strategic Implications*, edited by R. Fairhurst, K. Wall, and D. Cortright. Chicago, IL: Chicago University Press.

Wendt, Alexander. 1999. *Social Theory of International Politics*. Cambridge: Cambridge University Press.

Werner, Wouter G. 2004. 'From Justis Hostis to Rogue State: The Concept of the Enemy in International Legal Thinking'. *International Journal for the Semiotics of Law* 17: 155–68.

West, Lindy. 2015. 'The Real American Sniper Was a Hate-Filled Killer: Why Are Simplistic Patriots Treating Him as a Hero?' Accessed 12 April 2015. http://www.theguardian.com/commentisfree/2015/jan/06/real-american-sniper-hate-filled-killer-why-patriots-calling-hero-chris-kyle.

Westwood, David. 2005. 'Snipers'. In *The Encyclopaedia of World War I: A Political, Social, and Military History*, edited by S. C. Tucker. Santa Barbara, CA: ABC-CLIO.

Wheaton, Henry. 1836. *Elements of International Law: With a Sketch of the History of the Science*. Philadelphia, PA: Carey, Lea and Blanchard.

Wheeler, Everett L. 1987. 'Ephorus and the Prohibition of Missiles'. *Transactions of the American Philological Association* 117: 157–82.

Wheeler, Everett L. 1990. 'Review of The Western Way of War: Infantry Battle in Classical Greece by Victor Davis Hanson'. *The Journal of Interdisciplinary History* 21 (1): 122–5.

Wheeler, Nicholas J. 2000. *Saving Strangers: Humanitarian Intervention in International Society*. New York: Oxford University Press.

Wheeler, Nicholas J. 2002. 'Dying for Enduring Freedom': Accepting Responsibility for Civilian Casualties in the War against Terrorism'. *International Relations* 16 (2): 205–25.

Wheeler, Nicholas J. 2004. 'The Kosovo Bombing Campaign'. In *The Politics of International Law*, edited by C. Reus-Smit. New York: Cambridge University Press.

Whetham, David. 2005. *Just Wars and Moral Victories: Surprise, Deception and the Normative Framework of European War in the Later Middle Ages*. Boston, MA: Brill.

Whetham, David. 2013. 'Drones and Targeted Killing: Angels or Assassins?' In *Killing by Remote Control: The Ethics of an Unmanned Military*, edited by B. J. Strawser. Oxford: Oxford University Press.

Whiting, Marvin C. 2002. *Imperial Chinese Military History: 8000 BC–1912 AD*. New York: Writers Club Press.

Whittle, Richard. 2014. 'How We Missed Mullah Omar: An Inside Account of America's Botched First Predator Mission'. Accessed 10 October 2015. http://www.politico.com/magazine/story/2014/09/how-we-missed-mullah-omar-111026.

Wight, Martin. 1966. 'Why Is There No International Theory?' In *Diplomatic Investigations*, edited by H. Butterfield and M. Wight. London: Allen & Unwin.

Wikileaks. 2006. 'Terror Trials Update: Defendants Released and Rumours Unconfirmed'. Accessed 11 May 2017. https://wikileaks.org/plusd/cables/06SANAA448_a.html.

Wilcox, L. B. 2015. *Bodies of Violence*. Oxford: Oxford University Press.

Wilkeson, Frank. 1886. *Turned Inside Out: Recollections of a Private Soldier in the Army of the Potomac*. Lincoln, NE: University of Nebraska Press.

Will, George F. 2010. 'An NCO Recognises a Flawed Afghanistan Strategy'. Accessed 11 June 2017. http://www.washingtonpost.com/wp-dyn/content/article/2010/06/18/AR2010061803760.html.

Williams, Brian Glyn. 2010. 'The CIA's Covert Predator Drone War in Pakistan, 2004–2010: The History of an Assassination'. *Studies in Conflict and Terrorism* 33 (10): 871–92.

Williams, Jody. 2011. 'Borderless Battlefield: The CIA, the US Military, and Drones'. *International Journal of Intelligence Ethics* 2 (1): 2–34.

Williams, John. 2008. 'Space, Scale and Just War: Meeting the Challenge of Humanitarian Intervention and Trans-National Terrorism'. *Review of International Studies* 34 (4): 581–600.

Williams, John. 2015. 'Distant Intimacy: Space, Drones, and Just War'. *Ethics and International Affairs* 29 (1): 93–110.

Wilson, Christopher K. 1988. 'Marks of Honor and Death: Sharpshooter and the Peninsular Campaign of 1862'. In *Winslow Homer: Painting of the Civil War*, edited by M. Simpson. San Francisco, CA: The Fine Arts Museums of San Francisco.

Winner, Langdon. 1977. *Autonomous Technology: Technics-Out-of-Control as a Theme in Political Thought*. Cambridge, MA: MIT Press.

Witt, John Fabian. 2012. *Lincoln's Code: The Laws of War in American History*. New York: Free Press.

Wood, Graeme. 2015. 'Why We Fear and Admire the Military Sniper'. Accessed 11 April 2015. https://www.bostonglobe.com/ideas/2015/01/16/why-fear-and-admire-military-sniper/UKotOxOc2hTKoH9rZ5Kh9I/story.html.

Woods, Chris. 2011. 'Ten Years Since First Deadly Drone Strikes, Industry Gathers in London'. Accessed 3 November 2015. https://www.thebureauinvestigates.com/2011/11/21/drone-manufacturers-in-london-on-10th-anniversary-of-1st-strike/.

Woods, Chris. 2015. *Sudden Justice: America's Secret Drone Wars*. London: Hurst and Company.

Woollacott, Martin. 1999. 'A Sort of Peace: Smashing Serbia Was Not Pretty'. *The Guardian*, 4 June, 19.

Wragg, David W. 1986. *The Offensive Weapon: The Strategy of Bombing*. London: Hale.

Yenne, Bill. 2004. *Attack of the Drones: A History of Unmanned Aerial Combat*. Minneapolis, MN: Zenith Press.

Yeoman, Robert. 2015. 'Letters from Gallipoli'. Accessed 13 June 2017. http://www.irishtimes.com/culture/heritage/letters-from-gallipoli-1.2187873.

Zehfuss, Maja. 2011. 'Targeting: Precision and the Production of Ethics'. *European Journal of International Relations* 17 (3): 543–66.

Zenko, Micah. 2013a. 'Reforming U.S. Drone Strike Policies'. *Council of Foreign Relations Special Report* 65: 1–35.

Zenko, Micah. 2013b. 'If Trayvon Were Pakistani: Why Isn't Obama Outraged About a Drone War Based on Profiling?' Accessed 22 May 2017. http://foreignpolicy.com/2013/07/22/if-trayvon-were-pakistani/.

Zenko, Micah and Sarah Kreps. 2014. 'Limiting Armed Drone Proliferation'. Council on Foreign Relations: Centre for Preventive Action. Accessed 4 April 2019. https://cfrd8-files.cfr.org/sites/default/files/pdf/2014/06/Limiting_Armed_Drone_Proliferation_CSR69.pdf.

Ziegler, Charles A. 1994. 'Weapons Development in Context: The Case of the World War I Balloon Bomber'. *Technology and Culture* 35 (4): 750–67.

Zohar, Noam. 1993. 'Collective War and Individualistic Ethics: Against the Conscription of "Self-Defense"'. *Political Theory* 21 (4): 606–22.

Zohar, Noam. 2004. 'Innocence and Complex Threats: Upholding the War Ethic and the Condemnation of Terrorism'. *Ethics* 114 (4): 734–51.

Zurcher, Anthony. 2015. 'American Sniper: Was Chris Kyle Really a Hero?' Accessed 7 April 2019. https://www.bbc.com/news/world-us-canada-30923038.

Index

For the benefit of digital users, indexed terms that span two pages (e.g., 52–53) may, on occasion, appear on only one of those pages.

The manufacturer's authorised representative in the EU for product safety is Oxford University Press España S.A. of el Parque Empresarial San Fernando de Henares, Avenida de Castilla, 2 – 28830 Madrid (www.oup.es/en or product.safety@oup.com). OUP España S.A. also acts as importer into Spain of products made by the manufacturer.

www.ingramcontent.com/pod-product-compliance
Lightning Source LLC
LaVergne TN
LVHW011819200926
848484LV00005B/1248
* 9 7 8 0 1 9 8 8 5 1 4 6 2 *